Sustainable Mobility for Island Destinations

Theocharis Tsoutsos

Editor

Sustainable Mobility
for Island Destinations

 Springer

Editor
Theocharis Tsoutsos
School of Chemical and Environmental Engineering
Technical University of Crete
Chania, Greece

ISBN 978-3-030-86413-2   ISBN 978-3-030-73715-3   (eBook)
https://doi.org/10.1007/978-3-030-73715-3

Cover Photo identity: City of Rethymno, Crete. ©Renewable and Sustainable Energy Systems Lab-Technical University of Crete, Michael Pagalakis

This Springer imprint is published by the registered company Springer Nature Switzerland AG
The registered company address is: Gewerbestrasse 11, 6330 Cham, Switzerland

Editorial Preface

European cities are facing significant challenges related to mobility, in terms of congestion, air and noise pollution, fuel consumption, road accidents, accessibility, and quality of life. Nowadays, road transport accounts for 82% of citizens' personal transportation (EEA 2018), including cars, two-wheeled vehicles. and buses/coaches, while 49% of freight transport activity is road-related. Road transport vehicles are mostly using fossil fuels, which emit significant quantities of Greenhouse Gases (GHGs) and it is worth mentioning that nearly two-thirds of these emissions originate from light-duty vehicles. Road transport is the second biggest energy-consuming sector in the EU-28, after the residential sector and accounts for approximately 73% of the total transport GHGs emissions in EU (EEA, 2020). As a result, urban transportation has become one of the crucial factors for achieving the goals of the 2015 Paris Agreement on climate change commitments.

The **European Green Deal** has placed the primary policy drivers in order to make Europe climate neutral in 2050 [1]. A plan exists to reduce 90% EU's greenhouse gas emissions for 2050, introducing a new legislative framework. Additionally, the latest European Directives, as well as the Paris Agreement, demonstrate clearly the short and medium tendencies (European Commission 2019).

In this book are presented lessons learned by the CIVITAS DESTINATIONS project, which makes a tangible link between mobility and tourism in urban areas; the project funded by the Horizon 2020 programme. The project integrates the tourist mobility needs and the associated fluctuation impacts in the design and implementation of mobility solutions in order to strengthen the sustainability, attractiveness, accessibility, and efficiency of transport services and infrastructure for both residents and guests in six island cities: Funchal (Madeira-Portugal), Las Palmas de Gran Canaria (Spain), Elba (Italy). Limassol (Cyprus), Rethymno (Crete-Greece), and Valletta (Malta),

- Sustainable Urban Mobility Plans for guests and citizens
- Toward a Green Deal in Transport
- Sustainable Tourism and Mobility
- Measuring Environmental Impact in cities
- Innovation, ITS and tourism destinations

- Insular experiences and challenges from transport and tourism
- Greening public transport
- Micromobility in the urban environment

This book contains experience from a highly experienced group in the topics of sustainable tourism, mobility services experts, universities, and smart solutions design companies assisting: (a) the change of the mindset in insular and tourism areas; (b) the adoption of green mobility systems and services; (c) monitoring the environmental benefits to assist toward the Climate Change.

It includes analysis of implemented measures, useful examples, graphs, tables, as well as pictures. The organization of the book is as follows:

In the beginning, C. Mantero overviews the vision of the CIVITAS DESTINA-TIONS H2020 project, which has addressed the main challenges of tourism and resident mobility locally (rural, urban area, large city) and regionally, demonstrating 83 different sustainable mobility measures in six European islands (Malta, Cyprus, Elba, Crete, Gran Canaria, Madeira). The project findings focus on the quality of life and economic development to grant long-run environmental and financial sustainability of tourism and transport.

Afterward, E. Farmaki, M. Aryblia, S. Tournaki, and T. Tsoutsos present a multi-criteria assessment of 11 sustainable urban mobility measures according to 10 criteria for European medium-sized touristic cities. The study includes the viewpoint of six different European stakeholder groups, identifying their interests and comparing their ranking on appropriate mobility measures. This analysis can be a valuable tool for decision-makers during the shaping of future policies for sustainable mobility in urban tourist destinations, considering numerous parameters and stakeholders' viewpoint.

S. Gini and G. Ambrosino describe the role of the innovation and digital solutions played in facing the mobility needs of the rural areas and tourist destinations and in parallel to discuss the problems and conditions for their use in the provision of the different mobility and transport services. The innovation in the mobility provision chain is examined concerning the other relevant service elements (schemes, planning, programming, evaluation, and support) in order to clarify that innovation does not automatically mean digital solutions. Finally, besides the specific recommendations provided in each section, critical recommendations for Mobility Authority and Transport Operators are spotlighted.

After that C. Sessa illustrates a co-creation approach engaging stakeholders and citizens to develop a shared agenda of sustainable transport solutions, responding to travel needs which are qualitatively and quantitatively different for residents and tourists. This chapter shows the overall co-creation approach in all six DESTINA-TIONS demonstration sites, focuses on the application of the co-creation approach in the Elba island case—including the definition of a strategic agenda in a draft Sustainable Mobility Plan for the whole Elba island—and discusses lessons that can be drawn from the successes and failures.

Consequently S. Maas and M. Attard focus on the car-dependent island of Malta, which has increased congestion and rise to parking issues in many localities, taken-up limited space in the urban area, increased air and noise pollution, and created accessibility problems for pedestrians and cyclists. A survey was conducted to understand the awareness and acceptance of shared mobility services (bicycle, car, scooter sharing), by Maltese residents. Insights used to understand user needs and perceptions in light of such mobility innovations and what might encourage people to start using them as an alternative to private car use.

D. Tarasi, T. Daras, T. Tsoutsos are exploiting the opinion of the society in a typical Mediterranean city, Crete, identifying the mechanisms behind utility cycling and determining the aspects of using the bike in cities for daily commuting. This chapter focuses on the assessment of biking as an alternative mode of transport in a car-dependent city. Based on the opinion of the local society, high priority policies are recommended, such as proper infrastructure, sensitization campaigns, education programs.

Next I. Briesner highlights the rising importance of ecological tourism, which demands new perspectives of the tourist destinations in establishing new sustainable mobility structures and strategies for supporting regional economic development. She recommends Mobility Centers 2.0 as an efficient tool to reduce individual car use and the negative impact of visitor's travel in tourist regions, as well as to upgrade the quality of the leisure offer and the external image of the regions.

Finally, M. Aryblia, L. Quintal, M. Ribeiro, N. Sifakis, S. Tournaki, and T. Tsoutsos show the environmental monitoring systems implemented two different to measure and calculate a predetermined set of indicators, capable of estimating the environmental benefits in transport, economy, society, energy, and the environment in Funchal (Madeira-Portugal) and Rethymno, Crete-Greece. The monitoring systems were installed in specific sites all around these two cities, aiming, through the integrated sensors, to collect environmental data related to transport loads, such as environmental indexes (temperature, humidity, noise) and air pollutants (CO_2, CO, NO_x, SO_2, PM). The collection of critical and reliable data offers the opportunity of an adequate evaluation of the overall performance of the implemented measures toward sustainable, environmentally friendly, and low carbon mobility policies.

The greening of the islands is crucial for building a more sustainable future and a competitive local economy. The development of sustainable mobility in the islands and ecologically sensitive areas is directly related to the improvement of environmental quality. The need for preservation of their unique natural characteristics that are related not only to the natural but also to the cultural heritage, as well as the islands' limited carrying capacity, raise the issues of overloading and "flooding" by the related activities.

Furthermore, several islands consist of a major global tourism destination, so the potential minimization of their carbon footprint is essential for the success of the Paris Agreement. However, Insularity means isolation resulting in significantly highest transport costs, communications, and energy, when compared to the nearest continental regions; islands face several challenges to cover their green mobility transition:

- Several islands, as touristic destinations feature a high pressure from tourism and need to provide the high quality, sustainable environments desired by tourists;
- The seasonal fluctuation in their operation (winter/summer) impacts to the design of their necessary infrastructure (ports, roads, energy supply systems)
- Space use, fuel use, road safety, air quality, and noise pollution are issues that islands have to cope with.
- The size of most islands is missing the critical size to exploit the economy of scale to minimize the planning and installation cost.
- Though very often, the local/regional politicians are favorable, the transition has to face the current complicated local market context

Concluding, it obvious now that islands hold great opportunities. They are ideal test labs to develop/implement innovative systems and policies. On behalf of all respectful authors and contributors, I hope that this open publication will help the potential readers to understand better the above challenging topics.

Chania, Greece   Theocharis Tsoutsos

Acknowledgements The editor would like to express his gratitude to the authors who submitted chapters to this book, as well as all referees, who spent their valuable time providing their detailed reviews. Also to acknowledge his lab team members Maria Frangou and Stavroula Tournaki for their contribution during the development and review process of this book. This publication is supported by the European Commission by the European Union's Horizon 2020 research and innovation program, within the framework of the project CIVITAS DESTINATIONS (Project ID: 689031). The sole responsibility for the content of this paper lies with the authors. It does not necessarily reflect the opinion of the European Union. Neither the CINEA nor the European Commission is responsible for any use that may be made of the information contained therein.

References

European Commission (2019). The European Green Deal, 11.12.2019 COM(2019). Accessed from https://eur-lex.europa.eu/resource.html?uri=cellar:b828d165-1c22-11ea-8c1f-01aa75ed7 1a1.0002.02/DOC_1&format=PDF.
European Commission, DG MOVE (2020). Sustainable transport Accessed from https://ec.europa. eu/transport/themes/sustainable_en.
European Environmental Agency (2020). Progress of EU transport towards its environment and climate objectives Accessed from https://www.eea.europa.eu/themes/transport/term/term-bri efing-2018.
European Environmental Agency (2019). Greenhouse gas emissions from transport in Europe. Accessed from https://www.eea.europa.eu/data-and-maps/indicators/transport-emissions-of-greenhouse-gases/transport-emissions-of-greenhouse-gases-12.

Contents

Sustainable, Smart and Safe Mobility at the Core of Sustainable Tourism in Six European Islands

Claudio Mantero

Abstract Transport and tourism are rising economic sectors based on their mutual growth on reliable technological tools, affordable energy sources in relatively peaceful decades. This growing trend, faced along years of sudden slowdown caused mainly by the financial and health crisis; one of the most severe and recent episodes was the COVID-19 pandemic. The pandemic outbreak is representing a global turning point for both international markets of tourism and transport which can be addressed with innovative solutions looking to retake the growing trends. In such a changing framework CIVITAS DESTINATIONS European project addressed the main challenges of tourism and resident mobility at local (rural, urban area, large city) and regional levels, focusing on six European islands (Malta, Cyprus, Elba, Crete, Gran Canaria, Madeira) aiming to implement and evaluate 83 different sustainable mobility measures. The project proves to have a cumulative relevant impact on environmental indicators such as CO_2 emission avoided, fossil fuel saved and energy saved. Additionally, the project findings focus on the quality of life and economic development to grant long-run environmental and financial sustainability of tourism and transport at the local and regional levels. The main aspects observed were related to public participation and governance models in touristic mobility integrated with the residents' needs, touristic mobility the island, role of technologies to focus on final user needs, growing trends of elder and disabled people, new integrated and innovative business in tourism and mobility markets, health at the core of future trends.

Keywords Sustainable mobility · Green tourism · Insular economies · COVID-19 · Tourism value chain

C. Mantero (✉)
Horarios do Funchal, Transportes Publicos S.A., Madeira, Portugal
e-mail: ClaudioMantero@horariosdofunchal.pt

© The Author(s) 2022
T. Tsoutsos (ed.), *Sustainable Mobility for Island Destinations*,
https://doi.org/10.1007/978-3-030-73715-3_1

1

1 Introduction

Transport and tourism sectors are strictly related and interconnected with mutual growth dynamics. Tourist trends raised continuously and steadily along the last decades, reflecting the demand's stimulation coupled with pushing economies (Cristea & Mantero, 2020). Such trend drove the expansion of both tourism and transport markets which achieved higher figures and better performances year after year, as seen in Figs. 1 and 2.

The tourism sector has a key role at the European level, representing in 2019 near 10% of the total GDP. Focusing on the 6 countries that took part in the CIVITAS DESTINATIONS project through their insular regions, the figures of the Tourism sector for the country GPD is significant, varying from 11.6% in Malta to a maximum contribution of 22.66% in Cyprus (Pratt, 2015).

In CIVITAS DESTINATIONS, insular regions show a prominent role in the tourism industry compared to national economies. As possible to notice in Fig. 3, nearly 25% of the insular economies of Malta and Cyprus are related to the tourism sector, when on average national economies of the six sites rely on tourism only around 10% (WTTC, 2020).

Since 1950, both sectors' economies started to grow and faced along the years four great recessions period related to two financial crises 2001–2002 and recently in 2007–2008, a health crisis of SARS pandemic in 2003 and COVID-19 in 2020. The last decline due to COVID-19 affected enormously tourism and transport markets with an estimated deterioration of international tourism demand of 60%. Such a figure

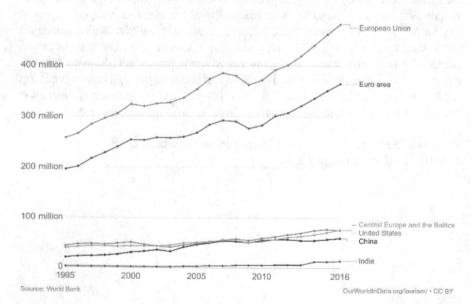

Fig. 1 International tourism demand—number of arrivals (Our world Data.org based on Word bank data)

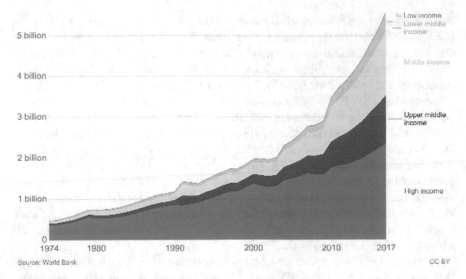

Fig. 2 Air transport demand—number of passengers (Our world Data.org based on Word bank data)

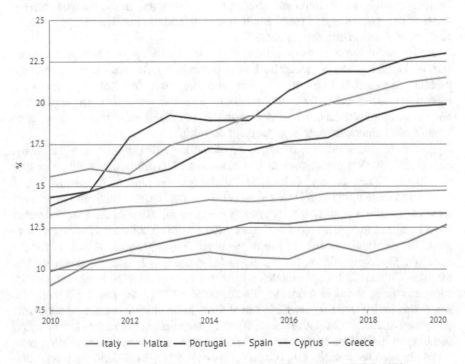

Fig. 3 Tourism sector weight in the GDP in DESTINATIONS countries (WTTC, 2020)

was estimated to rise to 80% until December 2020 if the pandemic recovery will delay (OECD, 2020). An average decline of transport in urban areas during the lockdown reached 75% and stabilized during July 2020 between 20 and 50%.[1] Despite the COVID-19 demand and offer downfall, recovery for both markets to pre-crisis figures is not expected before 2022 (OECD, 2020; IATA, 2020; Higgins-Desbiolles, 2020).

Before the COVID-19 outbreak, the trends of both industries shaped relevant positive effects at the economic and social levels, creating employment and improving national and local economic development. Tourism and travel represented in 2019, 10.3% of the global GDP with 330 million jobs created across the world representing the sector growths more after ICT and financial sectors (WTTC, 2020). Together with great benefits, tourism and travel industries generate relevant external adverse effects at social and environmental levels affecting the quality of life of cities (Koens et al., 2018), especially during the peak seasons. In such a general growing trend, the total balance is in almost all cases highly positive with more total benefits than private and social costs (Biondo, 2012).

After the COVID-19 pandemic outbreak in February/March, both sectors of transport and tourism started to recover the demand's decrease in June 2020. However, this period was considered by all stakeholders the right time to reassess the actual business models and to improve outdated processes based on the obsolete prerequisite of the 'golden age of travel and tourism', to address new rising challenges in customers' demand and rising market.

A successful business model, underneath the golden age of travel and tourism (UNWTO, 2016), can be traced back to two mobility milestones that changed the mobility of people dramatically for short and long distances. Both milestones are related to mass transport technological solutions and large distribution with robust and virtually infinite business models based on affordable resources (energy, land, urban spaces and manpower) and peaceful decades.

Affordable transport is a consequence of the large distribution of private cars, changing people's approach to distances and accessibility, introducing and promoting new lifestyles, new concept of time, new social relationships, more new habits and behaviours in general (MIND SETS, 2017). Transport modes modified the concept itself of vacation which was associated to the large diffusion of the family car and the principle of maximum mobility independence with no consideration for environment, liveability and quality of life of the touristic destinations, especially for the residents. During the fifties, the vacation idea resulted as a natural consequence of the consolidated car lifestyle model, with city planning oriented to the car and the socio-economic status of a private vehicle. More recently, the jumbo jet made the world smaller by enabling people to travel further and at less cost. Low-cost airline companies made air travel affordable, affecting the tourism experience and vacation planning in itself. Finally, the internet, mobile apps and social media, sharing platforms changed the holiday value chain (ITC—UNWTO, 2015), and the phases that build the touristic experience, including preparation, vacation and return.

---

[1] Data collected from in CIVITAS DESTINATIONS sites using automatic traffic counting data systems.

It is a fact that the constant rising trend of both industries was based on a blissful combination of affordable transport in mostly deregulated markets, incomes increasing, low cost of commodities and worldwide safe and secure travel conditions. In the light of a sudden global event such as COVID-19, both markets realize their urgent need to develop more resilient strategies to have the capacity to rethink the actual business model including innovative approaches to address the long-run human challenge of sustainable development defined globally by the UN sustainable development goals (Higgins-Desbiolles, 2020). Both tourism and mobility can recover from the recession. However, they must define common innovative long-run strategic development objectives, combining progressively radical game changers/innovations such as, new stricter environmental regulation, intensive digitalization, accessibility for elder and disable, integrated and multimodal transport modes for all (Mckinsey, 2020; Romão, 2020; Page, 2009).

2 Methodology and Data Collection

CIVITAS DESTINATIONS sustainable mobility project, which started in September 2016, was the perfect laboratory to study and test innovative solutions to attract tourists to use less pollutant transport services and to appraise, innovative technologies and business dynamics looking deeply at new interaction at the local level between Tourists and Residents. The project tested 83 different mobility measures divided into 6 clusters during 4 years in six European islands, addressing mutual empowerment aspects of tourism and mobility, focusing also on positive and negative interactions at the city level. During the implementation, both sectors showed relevant aggregate effects on the local economy, well-being and quality of life of both residents and tourists. Fig. 4 presents the relevant project data, main targets and the partners involved in CIVITAS DESTINATIONS.

Following the CIVITAS evaluation process (Dziekan et al., 2013) depicted in Fig. 5, local evaluation managers collected in each site ex ante and ex post relevant indicators to monitor the progress and impact of each mobility measure to verify the achievement of main pre-set outcome indicators. The evaluation process it's depicted in the following diagram of methodology.

Evaluation data set, collected at measure level, referred to Economy, Energy, Environment, Transport system and Society. In the project framework, it was possible to outline, at a macro level, three main clusters of measures: *shared mobility and walking, clean vehicles and public transport* and estimate the outcomes achieved. Tables 1, 2 and 3 present, respectively, for each cluster, relevant outcome indicators collected to assess key thematic areas in sustainable mobility.

Together with the quantitative evaluation, relevant qualitative information was collected and studied in specific qualitative process evaluation exercises focusing on hampering and enabling factors during the different measure implementation phases.

As a result, all insular sites showed clearly peculiar challenges in touristic mobility at destinations, such as the seasonality in managing transport services related to

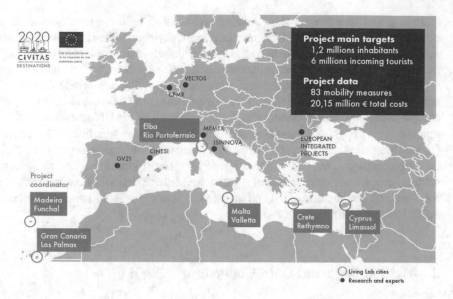

Fig. 4 The CIVITAS DESTINATIONS project sites, partners, main data and targets

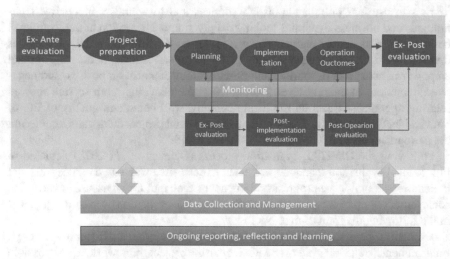

Fig. 5 Measure evaluation process applied in CIVITAS DESTINATIONS (Adaptation from Dziekan et al., 2013)

touristic demand, the growing trend of shared mobility resources together with walking options and lastly the growing spread of clean vehicle and public transport option (Gutiérrez & Miravet, 2016; Mantero & Gaudêncio, 2020).

At destination tourism sector interacts with transport services mostly as a neutral tool, one phase of the 'tourist value chain' process (ITC—UNWTO, 2015). The basic

Table 1 Project outcomes in shared mobility and walking cluster

Cluster of measures	Indicator	Outcomes
Las Palmas: LPA 4.1 Public e-bike system Limassol: LIM 4.2 Expansion of public bike-sharing system, including e-bikes	Average number of additional users (year)	13,813
	Number of additional bikes (including e-bikes)	507
	Additional km of cycle paths	199
	Fuel saved (litres/year)	188,855
	CO_2 emissions avoided (ton/year)	364

Table 2 Project outcomes in clean vehicles cluster

Cluster of measures	Indicator	Outcomes
Madeira: MAD 4.1 Promote the uptake of clean vehicles by fleet operators Limassol: LIM 4.1 Electric car rental connecting Limassol with airport and port Limassol: LIM4.3 Promote the uptake of electric vehicles, campaign on e-mobility Rethymno: RETH 4.1 Uptake of electric vehicles by fleet operators Elba: ELB 4.5 eV legislation revision and charging infrastructures	Percentage of additional electric vehicles (%)	50%
	Energy saved (MWh)	2,800
	Reduction in operating costs (€/year)	486,000
	Fuel saved (litres/year)	228,000
	CO_2 emissions avoided (ton/year)	774

Table 3 Project outcomes in public transport cluster

Cluster of measures	Indicator	Outcomes
Las Palmas LPA 7.2—Hybrid buses in the urban bus fleet Elba: ELB 7.1 Improve PT services for tourists Limassol: LIM 7.1 Improvement of PT routes, timetables, ticket procedure and bike transportation on buses to make the transport more attractive Malta: MAL 7.1 Integration of ferries into the Public Transport	Percentage of additional customers (year)	+10%
	Acceptance (%)	60%
	Fuel saved (litres/year)	10,200
	CO_2 emissions avoided (ton/year)	162

role of transport is to connect places and experiences to the tourists and in extreme and rare cases it can represent in itself the touristic experience such as toboggan as in Madeira Island, old vintage busses as the case of Malta or vintage trains (Lumsdon & Page, 2004). On the other side, transport sector developed during the last decade's specialized approach to tourism mobility demand with dedicated services such, tour operator airlines, charted busses, sightseen busses, car rental and touristic mobility options such tuk-tuk and other similar alternatives. Such touristic transport business can be associated with the blooming of new companies providing touristic mobility services providing innovative transport solution with exotic transport modes such rickshaw, tricycles, tuk-tuks, bike beers or in other cases such specialized touristic mobility services are a consequence of the vertical integration of tourism actors which expand their business sector including transport activities or internalizing transport companies (Grossman & Hart, 1986).

At a city level, it's clear that urban transport operators and services are mostly designed and planned according to the residents' needs except for specific lines or connections that have a specific touristic interest. Public transport, and the other transport schemes (i.e. DRT, ride-sharing) usually don't address specifically tourists as a core target passenger group for lacking data/information collection and capability/tool for monitoring the dynamic variations of the needs/demand. The 'simple' and common action carried out is the updating of the service programme on the basis of the season and main event. In the case of the mobility plan and evaluation process, usually, transport stakeholders collect limited information about tourist's mobility patterns with limited chance to define proper information and commercial strategies to tailor-made their services to touristic needs and attract more customers to sustainable transport option at touristic destinations.

3   Results and Discussion

In CIVITAS DESTINATIONS after the implementation of 83 urban and regional mobility measures in touristic destinations sites, with specific attention to tourist needs, mobility and tourism stakeholders have been following and responding to common main trends along with the project development, testing innovative solution and learning relevant lessons.[2]

3.1   Relevance of Decision Process and Data Collection to Act in Touristic Mobility

Stakeholder involvement and participated decision is essential in both touristic and mobility sectors to grant the acceptance of people and stakeholders (Kernel, 2005). Mutual exchange of information and dataset between both sectors is essential to improve the decision process at the city level reducing information overlapping and can bring to the definition of informed and agreed plan to address common issues affecting tourism and mobility such as traffic congestion. Tourism actors' contributions are essential to define a comprehensive mobility strategy at the urban level.

---

[2] Each measure has been duly assessed with ex ante and ex post evaluation methodology finish up a measure evaluation report (MER) document. All MERs will be release in final version on first semester 2021.

Fig. 6 Participatory in Portoferraio involving mobility and tourism stakeholders

As shown in Fig. 6, in Portoferraio, specific co-creation sessions brought relevant contributions to shape mobility and also to define long-run urban mobility strategy.

At the city level, transport and tourism sectors must share data to avoid the double work, to take advantage of synergies, to understand tourist's mobility patterns and to act in the more congested areas with specific measures. Transport technician's knowledge about touristic mobility is usually limited, however, this information is essential to define clear mobility plans at the urban level, for medium-long-run improvement of the quality of life of both residents and visitors.

3.2 Tourists at Destinations Prefer to Walk

In each site, it was applied a questionnaire to know tourist mobility to a representative sample of at least 400 tourists leaving the islands after the vacation experience. The data presented in Fig. 7 were collected at airports, ports and main touristic spots and are relevant to understand touristic mobility habits at a quantitative and qualitative levels.

Tourist with pre-booked transport mode options in the 6 site represents the great majority of tourists arrived at the destinations (48%). The transport choice is mainly oriented to cars, especially in places where airports are far from the main cities and public transport services are scarce or not integrated. Nevertheless, during the stay, an average of 33% of tourists, prefer to walk to move around and visit the destination.

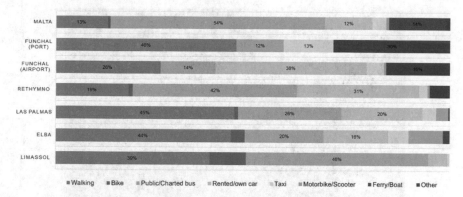

Fig. 7 Touristic transport modal share in each DESTINATIONS site

3.3 Role of Technological Tools to Simplify User Experience and to Improve Decision-Making

CIVITAS DESTINATIONS project assessed carefully technological tools involved in mobility planning and information in six cities, defining three different ITS categories: traffic management, public transport service operation, and mobility platforms. The main challenges observed in each site were the lack of integrated, shared, standardized information databases about mobility and tourism. Together with this fragmented data precondition, technological solutions implemented by public and private mobility actors are assorted, tailor-made and in many cases difficult to be integrated into the perspective of the end-user (tourist or resident). Such characteristics made local ITS solutions in small and medium cities incomplete and not integrated, affecting negatively the decision process in mobility and tourism markets. Such market failure brought to evident a lack of efficiency and effectiveness in the overall mobility offer (Ambrosino et al., 2016; Schwalbe, 1999).

Mobility as a service platform represents a customer-oriented conceptual model to allow users to compare and purchase the best mobility solution via a mobile app with an intuitive interface (Zijlstra et al., 2020). To address final users' needs, a relevant technological key feature is a multi-service sharing/booking intuitive platform which combines mobility and transport solutions with touristic services in simple information and purchase apps. Such platforms must be designed keeping in mind touristic/mobility customer needs, making easy and intuitive the selection process, the booking and the payment itself. Multiservice platforms (MP) as described in Fig. 8, consider tourists at the core of the services adapting and combining different complimentary services according to the final user perspective. In the case of Elba, the Mobility Sharing Agency addressed such technological challenges and design and implement a sharing mobility platform to provide information, allow booking, service payment, data quality assessment of services. The large-scale test was delayed due to the COVID-19, in any case, the launch and implementation were held on June 2020 and extensive evaluation of the platform performances was delayed at 2021.

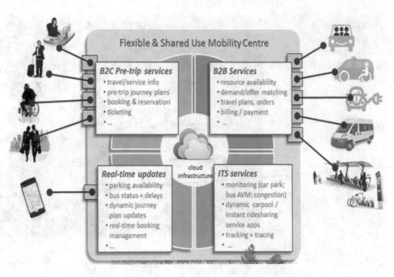

Fig. 8 The Shared Mobility Coordination Agency as kernel of Mobility as a Service (MaaS) transport offer system (Ambrosino, 2018)

3.4   Elder and Disabled People Growing 'Core Customer'

The baby boomer generation is one of the rising market segments in tourism. In 2017, in the EU, people with more than 55 years total up 38.5% of the total expenditure on tourist trips of at least one night. Such average data have relevant peaks considering the main outbound touristic markets of France (42.7%) and Germany (44.0%) (EUROSTAT, 2019). Together with the growing trend, disabled people's travel demand is also continuously growing in the EU with great potential for market specialization and innovation (EC, 2014). Considering such figures and trends accessibility and specific requirements dedicated to elder and disabled people are the core issues for the future of European tourism and sustainable mobility (EUROSTAT, 2019; UN, 2019). This trend represents not only an improvement challenge for already settled businesses and services but also an opportunity for new businesses or innovative tailored, on-demand services. Relevant measures were implemented in Funchal, Limassol, Rethymno and Portoferraio to improve accessibility for disable and visually and hearing-impaired people concerning walking facilities, public transport, informative tools, lighting systems, beaches access, transport hubs and figures were studied in Madeira in what regards public transport service adaptation to tourists needs in the specific case of elder and disabled people. Figure 9 depicts a measure implemented in Portoferraio to improve the accessibility of elder and disable tourists and residents.

Fig. 9 Accessibility measure implemented in Portoferraio to increase accessibility and reduce traffic speed

3.5 Mobility Franchises and Need for Tailor-Made Adaptation Case by Case

To have successful sustainable mobility systems, mobility franchises must be adapted to the local needs, habits and touristic patterns with relevant tailor-made interventions. Local businesses and local entrepreneurs are fully aware of local economic specifications and they play a key role to start local integrated services more attentive to unique tourist and resident needs. Such integrated and tailor-made solutions can be also upscaled in other regions with similar specifications and characteristics. Expressive findings of such a positive approach were evaluated in Las Palmas, Rethymno and Limassol for the bike-sharing systems.

3.6 Vertical and Horizontal Integration in Touristic Value Chain Market, Innovative Business Models and Market Regulation Challenges

In both sectors, and in recent years, it was verified a progressive improvement and transformation of the touristic value chain (ITC—UNWTO, 2015), and related supply chain. Mobility operators are expanding their market at the transport level merging services and providers. Such integration trend is not only visible in the same industry but also in complementary touristic industries such as leisure, travel agencies and accommodation services. Tourism operators, in a balanced way, are approaching transport services and providers, seeking integration opportunities, acting with an identical complementary aspiration in the mobility market. Such a trend in transport and tourism markets was proved in the six sites acting in the same industry or in other complementary industries in the tourism value chain process (Lafferty & Van

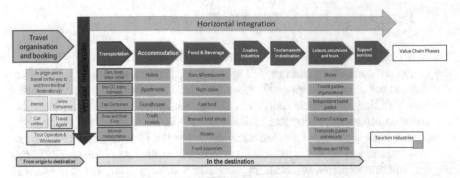

Fig. 10 Tourism value chain adaptation (ITC & UNWTO, 2015) to highlight the horizontal and vertical integration trends in mobility and tourism industries

Fossen, 2001). In the following figure, using the main reference of touristic value chain sequence is possible to identify the horizontal and vertical integration lines in Transport and Mobility sector at touristic destinations. In Fig. 10, vertical integration of different mobility operators in the same transport industry is depicted in red, while in yellow are highlighted the possible and growing integration of transport services with complementary touristic products/services such as tour operators, travel agencies, tourist guides, tourist attraction and others.

This situation leads to peculiar risks and opportunities. The main risks are related to the possible reduction of quality service in 'non-core' markets with fewer skills and experience and a potential oversupply product diversification in some cases cannot find enough demand to completely absorb the offer, with potential market saturation issue and the resulting need for better and improved regulation, to grant the minimum quality standards for residents and tourists. As a natural consequence of such integration, both tourist and mobility actors are keener and keener in working in wider markets involving complementary services or complementary industries to provide more sustainable services, not only at the touristic destination but also at the origin (Álvarez-Albelo & Hernández-Martín, 2012).

All in all, at the city level, touristic transport offer regulation is a key issue to avoid service overlapping and excessive congestion of main touristic areas and services providers. A clear and conservative market regulation of touristic mobility services is essential to drive a well-planned integration between touristic mobility and resident mobility needs. Market access, integration requirements and minimum service quality standards are key aspects that need to be studied and analysed at the local level with great attention, to speed up economic development with limited negative externalities.

3.7 Transport as Part of Touristic Experience

Transport service is mainly considered by the tourism sector as a way to deliver the touristic experience (Lumsdon & Page, 2004). The great majority of transport operators at the urban and regional levels are focused to provide high-level mobility services with limited touristic add value. This propensity is progressively changing and the mobility sector, at city and regional levels, started to be part of the touristic experience bringing enriching experiences to the tourists by itself or related to other services or products via the cross-selling or cross-discount tools (Ge, 2009). In this respect, relevant experiences about this successful strategy were tested in Madeira and Las Palmas de Gran Canaria with specific cross-selling/discount initiatives that combine transport and tourism services. In Madeira, Public Transport Friend initiative collect 27 touristic related shops and services providers which started to deliver cross discount to public transport passengers with a relevant increase of passengers within the tourism segment. In Las Palmas, a green loyalty system encourage citizens and tourists to adopt more sustainable lifestyle patterns by providing tangible economic rewards (points) using public transport. The points collected can be used to purchase products or services in shops, museums or theatres. In Fig. 11 is represented a schematic of the different measures implemented and tested in CIVITAS DESTINATIONS to increase the intrinsic value of transport as a tourism experience, moving from Transport as mere Utility to Transport as a Tourism experience.

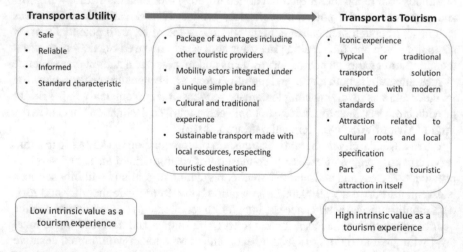

Fig. 11 Adaptation from Tourism and Transport, p. 7, Lumsdon and Page (2004) to present the measures tested in the sites that increase the intrinsic value of transport as a touristic experience

3.8 Health as a Cornerstone to Design Future Tourism and Mobility Services

According to EUROSTAT, data health issues represent one of the main reasons people aged 65 years and more in the EU do not travel (EUROSTAT, 2019). Such figures were affected in the light of the recent pandemic of COVID-19. The health aspect since March 2020 will be more and more at the core of the service supply design, and it is considered by tourism providers and potential customers a relevant issue to address with specific resources and skills. It's a fact that social distance, protection and hygiene will be considered during the pandemic period as part of the minimum quality standard for sustainable mobility services such as public transport and sharing services. Such measures were also been enforced in the sites with specific regulations tools to support health measures (Fig. 12) and to promote touristic destinations as safe and clean places via certifications and seal of disinfection.

Fig. 12 Information and communication campaigns examples in Madeira

4   Conclusions

Tourism and travel are key industries that globally represent in 2019, 10.3% of the total world GDP (WTTC, 2020). In insular contexts the role of both economies is pertinent (PRATT, 2015), in particular in the CIVITAS DESTINATIONS insular regions Tourism alone represents a core economy, reaching approximately 25% of the total island GDP (WTTC, World Travel and Tourism Council Data).

Considering the relevance of both sectors at the insular level, in the light of the recent challenges of COVID-19 which affected dramatically global tourism and travel demand, CIVITAS DESTINATIONS analyse the fundamentals of both economies, the interconnection of the industries and found innovative solutions to drive the upraise of the global demand of travel and tourism. In the six implementation sites, the project partners had the chance to develop and test during 4 years project effective measures to attract tourists to sustainable transport modes testing new business models, implementing alternative governance/regulation models and piloting integrated solutions. The main trends/main lessons about touristic mobility at destinations faced in the previous sections, can be summarised as the following ones: relevance of decision process and data collection to act in touristic mobility; tourists at destinations prefer to walk to discover places, the role of technological tools to simplify user experience to improve decision-making, elder and disable people growing 'core customer'; shared asset mobility franchises and need for tailor-made adaptation case by case; vertical and horizontal integration in touristic value chain market, innovative business models and market regulation challenges; transport as part of the touristic experience; health as the cornerstone to design tourism and mobility services.

The lessons and rising trends in tourism and transport depicted in this chapter want to contribute actively to feed the undergoing discussion about the future of urban mobility and tourism and the tourism value of chain addressing the challenge of seasonality and sustainable management of local resources to drive the innovation towards a more sustainable future for both industries. The measures tested in CIVITAS DESTINATIONS represent an effective way to address the challenges of sustainable mobility and tourism in the long run, addressing today's challenge of COVID-19 under the UN Goals references, bringing both sectors directly in a new 'silver age of tourism and transport' (EUROSTAT, 2019).

References

Álvarez-Albelo, C., & Hernández-Martín, R. (2012). Congestion and coordination problems in a tourism economy. *Tourism Economics, 18*(4), 691–710. https://doi.org/10.5367/te.2012.0136

Ambrosino, G., Nelson, J., & Gini, S. (2016). The pivotal role of public transport in smart mobility: The concept of flexible and shared mobility centre towards MaaS. In M. Schrenk, V. V. Popovich, P. Zeile, P. Elisei, & C. Beyer (Eds.), *REAL CORP 2016 Proceedings/Tagungsband*, 22–24 June 2016. http://www.corp.at. ISBN 978-3-9504173-0-2 (CD), 978-3-9504173-1-9 (print).

Ambrosino, G. (2018). *Shared use mobility agency in Elba island: From the concept to the IT Platform.* EU CIVITAS DESTINATIONS project. Retrieved September 07, 2020, from http://www.memexitaly.it/wp-content/uploads/2018/09/Booklet_Shared-Use-Mobility-Agency.pdf

Biondo, A. E. (2012). Tourism and sustainability: The positive externality. *International Journal of Sustainable Economy, 4*, 155–166.

Cristea, L., & Mantero, C. (2020, September). *Legacy document.* Internal working document in CIVITAS DESTINATIONS project.

Dziekan, E., Riedel, V., Müller, S., Abraham, A., Kettner, S., & Daubitz, S. (2013). *Evaluation matters, a practitioners' guide to sound evaluation for urban mobility measures.* Waxmann. Retrieved September 07, 2020, from https://www.researchgate.net/publication/304538182_Evaluation_matters_A_practitioners'_guide_to_sound_evaluation_for_urban_mobility_measures

European Commission. (2014). *Economic impact and travel patterns of accessible tourism in Europe*—Final report 2014. Retrieved September 07, 2020, from http://ec.europa.eu/DocsRoom/documents/7221/attachments/1/translations/en/renditions/native

Eurostat. (2019). *Ageing Europe, looking at the lives of older people in the EU* (2019 ed.).

Ge, Y. (2009). *The Sage handbook of tourism studies* (T. Jamal & M. Robinson, Eds., Chapter: 21). Sage Publications.

Grossman, S. J., & Hart, O. D. (1986). The costs and benefits of ownership: A theory of vertical and lateral integration. *Journal of Political Economy, 94*(4), 691–719. https://doi.org/10.1086/261404. ISSN 0022-3808.

Gutiérrez, A., & Miravet, D. (2016). *The determinants of tourist use of public transport at the destination in sustainability.*

Higgins-Desbiolles, F. (2020). The "war over tourism": Challenges to sustainable tourism in the tourism academy after COVID-19. *Journal of Sustainable Tourism.*

International Air Transport Association (IATA). (2020). *COVID-19 puts over half of 2020 passenger revenues at risk*, 14 April 2020. Retrieved September 07, 2020, from https://www.iata.org/en/pressroom/pr/2020-04-14-01/

International Trade Centre (ITC), & World Tourism Organization (UNWTO). (2015). *Tourism and trade: A global agenda for sustainable development.* Geneva.

Kernel, P. (2005). Creating and implementing a model for sustainable development in tourism enterprises. *Journal of Cleaner Production, 13*, 151–164.

Koens, K., Postma, A., & Papp, B. (2018). *Is overtourism overused? Understanding the impact of tourism in a city context.* November 2018 in Sustainability (ISSN 2071-1050; CODEN: SUSTDE).

Lafferty, G., & Van Fossen, A. (2001). Integrating the tourism industry: Problems and strategies. *Tourism Management, 22*(1), 11–19.

Lumsdon, L., & Page, S. J. (Eds.). (2004). *Tourism and transport: Issues and agenda for the new millennium.* Elsevier.

Mantero, C., & Gaudêncio. (2020). *Shaping mobility value chain in Madeira—The challenges of tourist mobility.* Traficom research reports 7/2020. Retrieved September 07, 2020, from https://www.traficom.fi/sites/default/files/media/publication/TRA2020-Book-of-Abstract-Traficom-research-publication.pdf

Mckinsey & Company. (2020). *The impact of COVID-19 on future mobility solutions.*

MIND SETS. (2017). EU cofunded project, Contract No. 640401, D. 3.2 *Future mobility challenges* and D 4.1 *New knowledge on mobility.*

Our world Data.org based on Word bank data. Retrieved September 07, 2020, from https://ourworldindata.org/tourism

Organisation for Economic Cooperation and Development (OECD). (2020). *Tourism Policy response to the coronavírus (COVID-19)*—Updated to 2nd June 2020. Retrieved September 07, 2020, from https://www.oecd.org/coronavirus/policy-responses/tourism-policy-responses-to-the-coronavirus-covid 19 6466aa20/

Page, S. J. (2009). *Transport and tourism: Global perspectives* (3rd ed.). Pearson Education

Pratt, S. (2015). The economic impact of tourism in Small Island Developing States. *Annals of Tourism Research, 52,* 148–160.

Romão, J. (2020). Smart specialization and tourism competitiveness. *Sustainability, 12*(14). Retrieved September 07, 2020, from https://doi.org/10.3390/su12145765

Schwalbe, U. (1999). *The core of economies with asymmetric information.* Springer

United Nation World Tourism Organization (UNWTO). (2016). *International tourist arrivals up 4% reach a record 1.2 billion in 2015.* Press release 16008. Retrieved September 07, 2020, from https://www.unwto.org/archive/global/press-release/2016-01-18/international-tourist-arrivals-4-reach-record-12-billion-2015

United Nations. (UN). (2019). Department of Economic and Social Affairs Population Division, World Population Ageing, New York, 2019.

World Travel and Tourism Council Data. (WTTC). (2020). Retrieved September 07, 2020, from https://wttc.org/Research/Economic-Impact

World Travel & tourism Council (WTTC). (2020). *Global economic impact & trends 2020*—June 2020. Retrieved September 07, 2020, from https://wttc.org/Research/Economic-Impact

Zijlstra, T., Durand, A., Hoogendoorn-Lanser, S., & Harms, L. (2020). Early adopters of mobility-as-a-service in the Netherlands. *Transport Policy, 97,* 197–209.

Assessing Sustainable Urban Mobility Policies in the Mediterranean Tourism Destinations Through Multi-Criteria Decision-Making Models

Eleni Farmaki, Maria Aryblia, Stavroula Tournaki, and Theocharis Tsoutsos

Abstract This chapter presents the assessment of 11 sustainable urban mobility measures according to 10 criteria for European medium-sized touristic cities, using multi-criteria decision-making. The study includes the viewpoint of six different European stakeholder groups, identifying their interests and comparing their ranking on appropriate mobility measures. It was found that the majority of stakeholders give the highest priority to the wellbeing of local communities and the quality of life, despite the economic implications of services and the potential impact on incoming tourism. Mostly they emphasise on at least two out of five criteria categories: Society and Environment or Society and Mobility. Tourism stakeholders showed a high preference for environmental criteria, demonstrating the continuously raising awareness on the links of tourism and environment. "Mobility management and travel plans" policy was the most popular policy amongst all groups, indicating that the provision of information, personalised plans, and smart applications can increase the use of sustainable mobility modes and have a significant positive impact in all examined categories. Overall, the multi-criteria analysis performed in this study can be a valuable tool for decision-makers during the shaping of future policies for sustainable mobility in urban tourist destinations, considering numerous parameters and stakeholders' viewpoints. Moreover, it can be further developed and adapted to specific needs.

Keywords Sustainable mobility solutions · Multi-criteria analysis · Decision-making · Sustainable urban planning

Abbreviations

MAMCA   Multi-actor multi-criteria analysis
MCDA   Multi-criteria decision analysis

E. Farmaki · M. Aryblia · S. Tournaki · T. Tsoutsos (✉)
Renewable and Sustainable Energy Systems Lab, School of Chemical and Environmental Engineering, Technical University of Crete, Chania, Greece
e-mail: theocharis.tsoutsos@enveng.tuc.gr

© The Author(s) 2022
T. Tsoutsos (ed.), *Sustainable Mobility for Island Destinations*,
https://doi.org/10.1007/978-3-030-73715-3_2

SUMP   Sustainable urban mobility plan

1 Introduction

Several studies stress the impact of tourism on local environmental and traffic conditions, resulting in significant air pollution, energy consumption (Page & Gue, 2016;
Aguiló et al., 2012), congestion (Saenz-de-miera & Rosselló, 2012), and road accidents; to alleviate these impacts, tourist destinations turn to the alternative transport
modes, such as public transport, shared vehicles, and high-quality services for visitors. (Gronau, 2017). The incorporation of the tourism aspect in the planning process
will allow tourist destinations to better address the fluctuation of demand and needs.

Sustainable Urban Mobility Plans (SUMPs) are designed, developed, and implemented having as central core the societies, not the transportation infrastructure,
aiming to provide a better quality of life within the urban locations. Integration,
participation, and evaluation processes are amongst the basic principles on which a
city can build its unique SUMP. Urban planners and decision-makers are often facing
numerous challenges, trying to integrate and gather previous knowledge, existing
successful practices, innovation, and new technology actions. The complexity level
of a decision-maker before the implementation of a SUMP is shown in Fig. 1, grouped
into four main steps:

1.   Preparation and analysis,
2.   Strategy development,
3.   Measure planning, and
4.   Implementation and monitoring (European Commission 2019).

Taking into account the rapid developments, the assessment of sustainable
mobility strategies is proven to be a complicated task, as it requires the examination
and incorporation of a wide variety of factors (environmental, social, technical, and
financial). A variety of techniques provide a flexible approach to deal with a wide
range of factors, thus providing significant assistance in decision-making, such as
Multi-Criteria Decision Analysis (MCDA) method.

Several studies have incorporated the interests of relevant stakeholders in the
multi-criteria assessment, focussing on sectors of planning, technical infrastructure.
Tsoutsos et al. (2009) investigated the sustainable energy planning for the island
of Crete using the multi-criteria analysis to assist the competent decision-makers
in defining the most suitable alternatives taking into consideration economical,
technical, environmental, and social criteria (Tsoutsos et al., 2009).

Macharis et al. (2010) developed the Multi-Actor Multi-Criteria Analysis
(MAMCA) methodology to evaluate transport projects (Tsoutsos et al., 2009), which
was applied to evaluate ten policies, in terms of mobility and logistics, in the "Flan-
ders in Action Process". The preferences of 11 stakeholder groups were represented

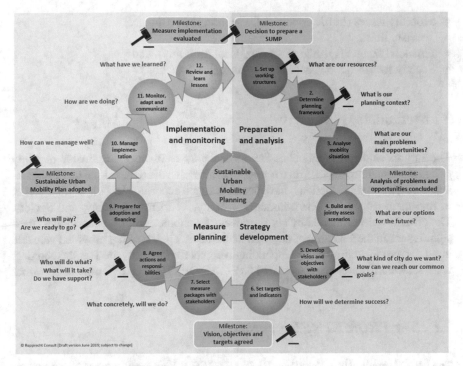

Fig. 1 The main steps of Sustainable Urban Mobility Plans (European Commission, 2019)

by evaluating pre-defined criteria. The most preferred policies were the stimulation of multimodal transport, the coordination of measures, and spatial planning.

Sun et al. (2015) evaluated six low-carbon transport policies in a Chinese city: tax and pricing adjustments, multi-operation mechanisms, environmental campaign, traffic demand management, and state funding and subsidies. Several stakeholders' groups were engaged in the process, and although their preferences varied, most groups considered state funding and subsidies as the most effective policy, along with traffic demand management.

Bulckaen et al. (2015) proposed a framework for ranking three small-scale urban and regional mobility projects that include policies different in theme, country, and objectives. The framework was a combination of MCDA to assess the sustainability of the projects and MAMCA to assess stakeholder preferences. The PROMETHEE method with 16 criteria was used, grouped under the three pillars of sustainability.

The main aims of this study are as follows:

- to offer a methodology to support the prioritising of existing mobility policies by the policymakers and
- to incorporate the various influence factors into the assessment of sustainable mobility policies for medium-sized, urban touristic areas of the Mediterranean.
- Environmental aspects (pollution, energy),

- Mobility issues (traffic conditions, transport infrastructure),
- Tourism flows,
- Financial factors (finance of service, local economy), and
- Society factors (safety, accessibility, users' satisfaction).

2 Methodology

The authors studied the influence factors in the assessment of sustainable mobility policies for medium-sized, urban touristic Mediterranean areas. A total of 11 sustainable mobility policies (actions) have been evaluated according to 10 specific criteria, in five main categories: Environment, Mobility, Tourism, Economy, and Society. The criteria were evaluated by six European stakeholder groups, according to their significance in the selection of appropriate policies. Their ranking provided valuable input for the calculation of corresponding weights for the analysis. The stakeholders were selected from the Horizon 2020 CIVITAS DESTINATIONS project network.

2.1 The PROMETHEE-Approach

The MCDA method employed was PROMETHEE, which ranks a set of n alternatives a_i ($i = 1, 2,\ldots, n$) based on a k criteria g_j ($j = 1, 2,\ldots, k$). Figure 2 presents the main steps of the PROMETHEE method, as followed in this study.

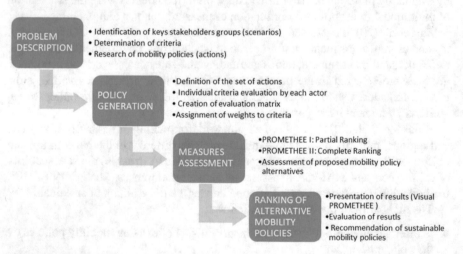

Fig. 2 Assessment Methodology of sustainable mobility policies

2.2 Sustainable Mobility Policies Alternatives

The selection of the specific strategies was based on the outcomes of related projects evaluating sustainable mobility policies. The goal was to assess a wide variety of mobility policies suitable for medium-sized tourist urban areas (Letnik et al., 2018), aiming to include eco-friendly transport solutions, current trends that promote social cohesion, recent technologies, solutions for infrastructure and equipment, "soft" policies, and tourist-oriented services. The selection of policies was also highly influenced by the sustainable mobility actions carried out in the frame of the CIVITAS DESTINATIONS project in the city of Rethymno. The selected mobility policies were adjusted to the specific objectives of the study, including most transport modes: car, public transport, and active mobility (cycling, walking).

The examined alternative sustainable mobility policies were as follows:

- Sustainable Urban Mobility and Logistic Plans,
- Smart metering systems,
- Increased traffic safety and security—Eco and safe driving training,
- Mobility plans for school communities,
- Attractive and accessible public spaces,
- Shared mobility services,
- E-charging infrastructures and e-vehicles in public fleets,
- Mobility management and travel plans,
- Behavioural change and informative actions,
- Low emission zones and parking management, and
- Improved and accessible PT services for tourists and residents.

2.3 Sustainable Mobility Criteria

Criteria were selected based on relevant researches and frameworks complied, to assess transport policies (Awasthi et al., 2018). The sustainable mobility measures (actions) were evaluated according to specific criteria, covering five main categories: Environment, Mobility, Tourism, Economy, and Society (Table 1).

2.4 Actors Involved in Sustainable Mobility Planning

Groups of stakeholders were mapped at first, assuring the inclusion of all key categories, in terms of demand and offer (i.e. users/operators); public and private experts (i.e. academics/consultants); and governance and non-profit organisations, according to the pillars of sustainable development (economy, environment, and society). The additional element in this specific study is the involvement of tourism actors, a

Table 1 Description of evaluation criteria

Category	Criterion	
Environment	Energy (ENE)	Reduction of energy/ fuel consumption and the share of conventional fuel in the implementation area
	Environmental pollution (ENV)	Reduction of average GHG emissions and noise levels in the implementation area
Mobility	Traffic conditions (TRC)	Modal share shift towards alternative transport of the target group involved and Traffic flow improved in the examined area
	Transport infrastructure (TRI)	Level of intermodal integration of transport services, along with existing infrastructure
Tourism	Tourist flow (TOU)	Increased share of tourists using transport services, No. of incoming tourists, and GDP generated by tourism
Economy	Service finance (SEF)	Cost of new services and infrastructure, including capital costs and maintenance
	Local economy (LOE)	Level of the increased affordability of public transport services for the users, level of financial gain by new services, and infrastructure for operators
Society	Safety (SAF)	Level of perceived road safety and security amongst target groups involved and reduction of No. of road incidents
	Users Satisfaction (USS)	Level of satisfaction and acceptance of the mobility policies amongst the target groups involved
	Accessibility (ACC)	Level of accessibility of transport services and infrastructure and perception of accessibility amongst users

significant segment of stakeholders, representing the specific needs and motivations of visitors as regards to mobility.

The participating actors, located in different European countries, mainly in Mediterranean touristic urban areas, including Greece, belonged to six groups:

(a)   Local Authorities (LAs),
(b)   Transport Operators and Mobility Providers (TOs),
(c)   Tourism Sector (TS),
(d)   Academic Institutions (AI),
(e)   Mobility Experts (ME), and
(f)   Environmental Groups (EG).

2.5 Evaluation Table

Based on extensive literature research and evaluation reports, the evaluation matrix was formed (Table 2). For the evaluation criteria, the values of the sustainable mobility alternative policies were ranked on a five-point scale.

2.6 Weight Factors

The selected criteria were evaluated in respect to their significance; their ranking provided the required data for the calculation of corresponding weights for the analysis. Each member of the panel completed a classification table, indicating their order of preference, from the most important criterion (1) to the least important (10). The relative weights were calculated for each response, while the weight of each criterion per stakeholder group was calculated as the average value of the relative weights of the actors in the specific group.

The mean weight of each criterion was calculated and the weights normalised, so that they become comparable (Tsoutsos et al., 2009).

2.7 Sensitivity Analysis

A sensitivity analysis was performed, using the tools of the Visual PROMETHEE software, such as "Walking Weights" that allows altering values of weight factors to observe their impact in the final classification and "Stability Intervals" that defines the alteration limits of weights inside which the final classification remains identical unchanged.

3 Results

3.1 Calculation of the Weight Factors

Following the aforementioned methodology per stakeholder group, relative weights (%) for every group were calculated (Table 3).

3.2 Ranking of the Alternative Policies Per Group

In Table 4 are shown the high policy priorities according to the groups.

Table 2 Evaluation table of sustainable mobility policies

Evaluation table		Criteria[a]									
		Environment		Mobility		Tourism	Economy		Society		
Actions		ENE	ENV	TRC	TRI	TOU	SEF	LOE	SAF	USS	ACC
Sustainable Urban Mobility and Logistic Plans	1	2	1	2	5	3	1	2	3	3	4
Smart metering systems/ Real-time mobility information	2	1	1	2	4	1	3	2	2	4	1
Increased traffic safety and security/ Eco-driving training	3	2	2	2	4	2	3	4	4	4	3
Mobility plans for school communities	4	2	2	2	3	1	5	2	3	4	4
Attractive and accessible public spaces	5	1	1	3	2	2	2	4	3	5	4
Shared mobility services (bike, car, taxi)	6	1	2	2	3	3	3	2	1	3	2

(continued)

Table 2 (continued)

Evaluation table	Criteria[a]									
Actions	Environment		Mobility		Tourism	Economy	Society			
	ENE	ENV	TRC	TRI	TOU	SEF	LOE	SAF	USS	ACC
7 E-vehicles in public fleets and e-charging infrastructures	2	3	1	2	2	3	2	1	3	1
8 Mobility management and travel plans	2	3	4	5	5	1	3	2	5	3
9 Behavioural change and informative actions	1	1	2	1	2	1	1	3	5	2
10 LEZs and parking management	3	5	5	2	1	1	2	3	3	2

(continued)

Table 2 (continued)

Evaluation table

Actions	Criteria[a]	Environment		Mobility		Tourism	Economy		Society		
		ENE	ENV	TRC	TRI	TOU	SEF	LOE	SAF	USS	ACC
Improved and accessible PT services for guests and residents	11	1	1	2	3	2	3	3	2	3	4
RANKING DESCRIPTION	5	Very high (40–50%)	Very high (40–50%)	Very high (40–50%)	Very high (all modes)	Very high (20–25%)	Very low (0–200,000 €)	Very high (20–25%)	Very high (80–100%)	Very high (80–100%)	Very high (80–100%)
	4	High (30–40%)	High (30–40%)	High (30–40%)	High (3–4 modes)	High (15–20%)	Low (200,000–400,000 €)	High (15–20%)	High (60–80%)	High (60–80%)	High (60–80%)
	3	Moderate (20–30%)	Moderate (20–30%)	Moderate (20–30%)	Average (2–3 modes)	Moderate (10–15%)	Moderate (400,000–600,000)	Moderate (10–15%)	Moderate (40–60%)	Moderate (40–60%)	Moderate (40–60%)
	2	Low (10–20%)	Low (10–20%)	Low (10–20%)	Low (1–2 mode)	Low (5–10%)	High (600,000–800,000)	Low (5–10%)	Low (20–40%)	Low (20–40%)	Low (20–40%)

(continued)

Table 2 (continued)

Evaluation table											
Actions	Criteria[a]										
	Environment		Mobility		Tourism	Economy	Society				
	ENE	ENV	TRC	TRI	TOU	SEF	LOE	SAF	USS	ACC	
1	Very low 0–10%	Very low 0–10%	Very low 0–10%	Very low (no mode)	Very low 0–5%	Very high (>800,000 €)	Very low 0–5%	Very low 0–20%	Very low 0–20%	Very low 0–20%	

[a]ENE: Energy, ENV: Environmental pollution, TRC: Traffic conditions, TRI: Transport infrastructure, TOU: Tourist flow, SEF: Service finance, LOE: Local economy, SAF: Safety, USS: Users satisfaction, ACC: Accessibility

Table 3 Weights (%) matrix for all stakeholder groups

Group	ENE	ENV	TRC	TRI	TOU	SEF	LOE	SAF	USS	ACC
Local Authorities (LA)	11.36	13.79	11.82	11.52	6.21	3.64	**3.48**	11.97	11.21	**15.00**
Transport Operators (TO)	7.27	9.82	**13.82**	12.00	6.55	8.55	11.45	12.18	**4.91**	13.45
Tourism Sector (TS)	15.76	**16.97**	8.79	10.30	9.70	**3.94**	7.58	9.70	6.36	10.91
Academic Institutions (AI)	6.36	15.00	11.36	7.73	7.27	**4.09**	10.91	**15.45**	9.09	12.73
Mobility Experts (ME)	8.64	9.85	12.27	10.76	10.76	**6.21**	7.73	8.33	12.27	**13.18**
Environmental Groups (EG)	14.55	**17.58**	**5.45**	10.91	6.06	10.00	6.97	11.52	6.67	10.30

Local Authorities (LA): the accessibility criterion evaluated as the most important with 15.00%, and the least important criteria were the economic ones

Transport Operators (TO): EU TOs showed a preference for technical and social criteria. The least important criterion was the tourist flow

Tourism Sector (TS): showed a preference for environmental criteria. The least important criteria were service finance (EU)

Academic Institutions (AI): the highest weight factor was safety; the least important is service finance

Mobility Experts (ME): EU MEs expressed a maximum preference for the accessibility criterion and the minimum for service finance

Environmental Groups (EG): EU EGs put greater emphasis on environmental criteria, giving the highest weight factors to environmental pollution and energy; the least important criterion was "traffic conditions" (5.45%)

LA: Local Authorities, TO: Transport Operators/ Provides / Services, TS: Tourist Sector, AI: Academic Institutions, ME: Mobility Experts, EG: Environmental Groups

3.3 Total Classification of Alternatives Solutions

According to the results of the PROMETHEE II Complete Ranking for all stakeholder groups (Fig. 3), "Mobility management and travel plans" policy is classified as the most suitable policy, followed by "Increased traffic safety and security—Eco-driving training" and the "LEZs and parking management" as a second option.

The medium-ranked policies for EU groups by descending order are as follows: Sustainable Mobility Plans, Mobility plans for school communities, Attractive and accessible public spaces, Shared mobility services, Improved and accessible PT services for tourists and residents, and Behavioural change and informative

Table 4 PROMETHEE II ranking for all EU stakeholder groups

Ranking table	Stakeholder group						Total
Actions	LA	TO	TS	AI	ME	EG	
Sustainable urban mobility plans/sustainable urban logistic plans	4	4	4	6	4	5	**4**
Smart metering systems/real-time mobility information	10	9	9	10	10	9	**10**
Increased traffic safety and security—eco-driving training	2	2	2	2	2	2	**2**
Mobility plans for school communities	5	5	5	5	6	4	**5**
Attractive and accessible public spaces	6	6	7	4	5	7	**6**
Shared mobility services (bike, car, taxi)	7	8	6	8	8	6	**7**
E-charging infrastructures and e-vehicles in public fleets	11	11	11	11	11	11	**11**
Mobility management and travel plans	1	1	1	1	1	1	**1**
Behavioural change and informative actions	9	10	10	9	9	10	**9**
Low emission zones and parking management	3(3)	3(3)	3(3)	3(3)	3(6)	3(3)	**3**
Improved and accessible PT services for tourists and residents	8	7	8	7	7	8	**8**

actions, while the second-worst policy was "Smart metering systems"; "E-charging infrastructures and e-vehicles in public fleets" policy was classified as the worst.

As observed, LA, TO, and TS present the exact same ranking for seven policies and slight differences for the remaining. The members of the groups AI, ME, and EG exhibit similarities between them and with the first three groups for 3–5 policies.

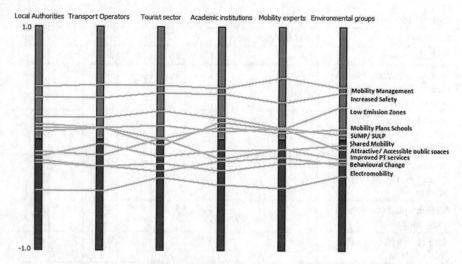

Fig. 3 Scenarios comparison PROMETHEE II ranking for all EU stakeholder groups

4 Discussion

4.1 Scenarios Comparison

Figures 2 and 3 depict the classification of PROMETHEE II for all stakeholder groups, comparing the different scenarios. Per group, vertical green and red lines are displayed (Phi+ = 1.0, Phi− = −1) and horizontal blue lines, representing each policy, cross the vertical lines, representing the value each action received, and attribute the net flow values (Phi) per criterion.

The "Mobility management and travel plans" policy was ranked as the most suitable amongst all groups, the "Increased traffic safety and security—Eco-driving training" policy as the second most proper, and the "E-charging infrastructures and e-vehicles in public fleets" as the worst one, due to their performance in highly weighted criteria (Fig. 3). It can be observed that the in-between ranking of the policies per group differs in a higher or lower degree.

4.2 Sensitivity Analysis

The abovementioned ranking of the sustainable mobility policies per group was finally determined by the weights given to the criteria; however, the weights' impact in the final classification is not clear. In order to identify whether the final ranking would change without the weight factors attributed, the "Walking weights" tool was used. Figure 4 presents the final ranking of the policies in case of equal weights,

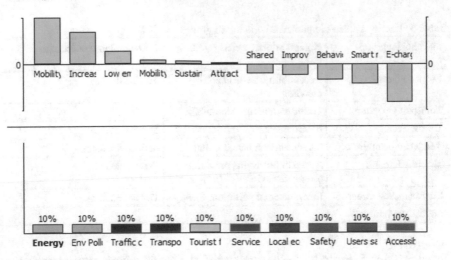

Fig. 4 Walking weights for sustainable mobility criteria

setting a 10% weight factor per criterion (= 100%/10 criteria).

It can be observed that "Mobility management and travel plans" and "Increased traffic safety and security—Eco-driving training" policies remain at the top of the PROMETHEE II ranking, the "E-charging infrastructures and e-vehicles in public fleets" policy remains at the bottom, while the ranking of the remaining policies changes for some groups.

Additional tools for sensitivity analysis were applied, as provided by the Visual PROMETHEE software, to assess the interval stability of the criteria. The "stability intervals" per criterion present the alteration limits in relation to the calculated weights; this analysis can be applied to all policies in order to examine the potential alteration of the total ranking or it can even be applied to the top-ranked policies to examine their stability. The stability interval of the weights for "Mobility management and travel plans", in reference to the weights, was calculated. The minimum values do not exceed 1.36%, while the maximum values of the weights range from 17.72% to 100%. The stability intervals indicate that the top-ranked policy is very stable. At the same time, the criteria "Energy" and "Safety" are the most "sensitive" at both levels, since they have the lowest alteration limits, meaning that if the weights exceed the maximum value, the "Mobility management and travel plans" would no longer be the top-ranked policy.

4.3 Identifying Stakeholders' Interests

Despite the fact that weight factors do not drastically change the ranking of the policies, they can provide valuable insights into the interests of each stakeholder group. According to the results of this process, the priority criteria per group are

Table 5 Priority and least important criteria for all groups

Stakeholder group	Highest priority criteria—EU level	Lowest important criteria—EU level
Local Authorities	Accessibility, environmental pollution	Service finance
Transport Operators	Traffic conditions, accessibility	Tourist flow
Tourist Sector	Environmental pollution, energy	Service finance
Academic Institutions	Safety, environmental pollution	Service finance
Mobility Experts	Accessibility, traffic conditions, users satisfaction	Service finance
Environmental Groups	Environmental pollution, Energy	Traffic conditions
Local Communities	–	–

listed in Table 5. It can be easily perceived that specific criteria have prevailed in the preferences of two or more groups, highlighting the interdependencies between groups' interests.

Most groups emphasise on at least two out of five criteria categories: Society and Environment or Society and Mobility, the most important criteria being Environmental pollution, Accessibility, Safety, Energy, and Traffic conditions.

Overall, two specific stakeholder groups present specific but anticipated priorities: EG gives priority strictly to environmental criteria, as expected, (Macharis et al., 2010, Lebeau et al., 2018, Bergqvist et al., 2015). However, TS also places the highest weights on environmental criteria (Michailidou et al., 2016), followed by social and tourism criteria, acknowledging the raised awareness on the links of tourism and environmental pollution (Abdul et al., 2017).

Amongst the least important criteria, service finance prevailed for most groups, indicating that economic demands would not determine the implementation of a potential mobility policy.

Overall, most groups give priority to criteria related to the wellbeing of local communities and the quality of life, despite the economic implications of services and the potential impact on incoming tourism.

5 Conclusions and Recommendations

This chapter aimed to assess 11 sustainable urban mobility policies for European medium-sized touristic cities in the Mediterranean region, integrating stakeholders' viewpoint through MCDA, more specifically with the use of the PROMETHEE model. As an added value, the study incorporated (a) the tourism aspects in various steps of the analysis and (b) a two-level stakeholder involvement approach and comparison of European and local stakeholders' interests and results.

The study revealed that most stakeholders give priority to the wellbeing of local communities and the quality of life, despite the economic implications of services and the potential impact on incoming tourism. Most EU groups emphasise on at least two out of five criteria categories: Society and Environment or Society and Mobility. Interestingly, the Tourism Sector group expressed a preference for environmental criteria, demonstrating the continuously raising awareness on the links of tourism and environment.

Overall, environmental pollution, accessibility, safety, energy, and traffic conditions were identified as the most important criteria. Service finance prevailed as the least important for most EU level groups.

For all groups, "Mobility management and travel plans" policy was classified as the most popular policy, indicating that the provision of information, personalised plans, and smart applications can increase the use of sustainable mobility modes and have a significant positive impact on all examined categories. On the other hand, the "E-charging infrastructures and e-vehicles in public fleets" policy was classified for all groups as the worst option, and despite being considered the "green" alternative to conventional vehicles, this policy doesn't have significant impacts on all aspects examined.

The analysis ranked and identified the optimal sustainable mobility policies, based on their overall performance to the weighted criteria. However, when the three top-ranked policies were further analysed according to their performance on priority criteria, it was found that although they present a good overall performance, one may overcome the other on specific criteria. Having this in mind, it can be assumed that the ranking consists of a suggestion and the final selection by a decision-maker can be different, according to targeted needs.

The incorporation of stakeholders' viewpoint is essential for an integrated approach and under this principle, this study took place, proposing the optimal solution for each stakeholder group according to their preferences. Nevertheless, the study could be further enhanced by the inclusion of a higher number of participating stakeholders, in order to have an even more representative sample of actors; especially the inclusion of local communities and users at the EU level would make the comparison of the two levels feasible.

Further research could include the addition of a time-related criterion, aiming to evaluate the preparation and implementation period per policy required to present potential impact and, in this way, identify short-term and long-term policies that might be more suitable according to the specific areas. On the other hand, the financial criteria could be omitted from the assessment, since—according to the results—they were considered of less significance. Nonetheless, if required, they can be further examined through a cost–benefit analysis. In cases where external funding sources are available, financial implications are less concerning for local authorities.

Further, the enhancement of the evaluation matrix with existing data related to a particular city/area or with real data collected/monitored in the city will strengthen the analysis, increasing the precision of the results and thus facilitating the shaping of future local policy. In addition, it would be particularly interesting for decision-makers to assess different sustainable mobility policies under the same thematic area (e.g. different safety policies or various smart metering systems), in case a specific area of intervention is already identified, or even proceed to such approach as a "second" stage analysis, following the suggestions of the initial analysis.

In terms of participation, the tourism sector and the local communities were particularly "difficult" teams to involve, a fact indicating the lack of awareness and/or active engagement and participation in mobility planning processes. Speedy answers were provided by academic institutions and mobility experts (29 answers). The necessary data gathering for the evaluation matrix fulfilment was faced by the selection of suitable data from already demonstration projects and relevant publications in order to achieve a high-quality and reliable estimation of performance.

As an overall conclusion, the multi-criteria analysis performed in this study can be a valuable tool for decision-makers during the shaping of future policies for sustainable mobility in urban tourist destinations, considering numerous parameters and stakeholders' viewpoints, but it can be further developed and adapted to specific needs.

References

Abdul, S., Khan, R., Qianli, D., Songbo, W., Zaman, K, & Zhang, Y. (2017). Travel and tourism competitiveness index : The impact of air transportation, railways transportation, travel and transport services on international inbound and outbound tourism. *Journal of Air Transport Management, 58*, 125–134.

Aguiló, E., Palmer, T., & Rosselló, J. (2012). Road transport for tourism: Evaluating policy measures from consumer profiles. *Tourism Economics, 18*(2), 281–293.

Awasthi, A., Omrani, H., & Gerber, P. (2018). Investigating ideal-solution based multicriteria decision making techniques for sustainability evaluation of urban mobility projects. *Transportation Research Part A: Policy and Practice, 116*, 247–259.

Bergqvist, R., Macharis, C., Meers, D., & Woxenius, J. (2015). Making hinterland transport more sustainable a multi actor multi criteria analysis. *Research in Transportation Business & Management, 14*, 80–89.

Bulckaen, J., Keseru, I., Donovan, C., Davies, H., & Macharis, C. (2015). Development of a new evaluation framework for urban and regional mobility projects. In *Benelux interuniversity association of transport researchers: transportation research days 2015.*

European Platform on Sustainable Urban Mobility Plans. (2019). *Guidelines for developing and implementing a sustainable urban mobility plan* (Second Edition). Accessed April 2020, from https://www.eltis.org/sites/default/files/guidelines_for_developing_and_implementing_a_sustainable_urban_mobility_plan_2nd_edition.pdf

Gronau, W. (2017). On the move: Emerging fields of transport research in urban tourism. In *Tourism in the city* (pp. 81–91). Springer, Cham.

Lebeau, P., Macharis, C., Van Mierlo, J., & Janjevic, M. (2018). Improving policy support in city logistics: The contributions of a multi-actor multi-criteria analysis. *Case Studies on Transport Policy, 6*(4), 554–563.

Letnik, T., Luppino, G., & Bardi, A. (2018). Review of policies and measures for sustainable and energy efficient urban transport. *Energy, 163*, 245–257.

Macharis, C., De Witte, A., & Turcksin, L. (2010). The multi-actor multi-criteria analysis (MAMCA) application in the Flemish long-term decision making process on mobility and logistics. *Transport Policy, 17*(5), 303–311.

Michailidou, A. V., Vlachokostas, C., & Moussiopoulos, N. (2016). Interactions between climate change and the tourism sector: Multiple-criteria decision analysis to assess mitigation and adaptation options in tourism areas. *Tourism Management, 55*, 1–12.

Page, S., & Gue, Y. (2016). Transportation and tourism: A symbiotic relationship. In T. Jamal & M. Robinson (Eds.), *The Sage Handbook of Tourism*. Sage Publications.

Saenz-de-Miera, O., & Rosselló, J. (2012). The responsibility of tourism in traffic congestion and hyper-congestion: A case study from Mallorca, Spain. *Tourism Management, 33*(2), 466–479.

Sun, H., Zhang, Y., Wang, Y., Li, L., & Sheng, Y. (2015). A social stakeholder support assessment of low-carbon transport policy based on multi-actor multi-criteria analysis: The case of Tianjin. *Transport Policy, 41*, 103–116.

Tsoutsos, T., Drandaki, M., Frantzeskaki, N., Iosifidis, E., & Kiosses, I. (2009). Sustainable energy planning by using multi-criteria analysis application in the island of Crete. *Energy Policy, 37*(5), 1587–1600.

Innovation, Digital Solutions and MaaS Concept in Touristic Insular and Rural Destinations: The Case of the Shared Mobility Agency in Elba

Saverio Gini and Giorgio Ambrosino

Abstract This chapter aims to describe the role of innovation and digital solutions in facing the mobility needs of the rural areas and tourist destinations and, at the same time, to discuss the problems and conditions for their use in the provision of the different mobility and transport services. Innovation in the mobility provision chain is analysed in relation to the relevant service elements (schemes, planning, programming, evaluation and support) in order to clarify that innovation does not automatically mean digital solutions. An overview of the emerging trends in digital solutions is provided, focusing on the Mobility as a Service (MaaS) concept and mobile applications. For the APP, market-specific considerations are developed on the different roles and typologies, making a classification for understanding current Mobility APP "Jungle". Moreover, the concept of Shared Use Mobility Agency (SUMA) is presented. SUMA aims to integrate smart mobility services under the concept of MaaS in order to answer to the mobility needs of residents and tourists acting on the demand and providing coordinated collective and ridesharing transport services. SUMA could be considered an "innovative" approach mixing advanced digital architecture and shared mobility schemes with the relevant organization, operation and business elements. Section 5 presents the SUMA implementation in Elba island (Regione Toscana-Italy), within CIVITAS DESTINATIONS project underlying the specifications of the platform and APP. Finally, besides the specific recommendations provided in each section, key recommendations for Mobility Authority and Transport Operators are spotlighted.

Keywords Ridesharing services · Info mobility · Innovation · ITS · Digitalization · MaaS

S. Gini · G. Ambrosino (✉)
MemEx srl, Livorno, Italy
e-mail: giorgio.ambrosino@memexitaly.it

S. Gini
e-mail: saverio.gini@memexitaly.it

© The Author(s) 2022
T. Tsoutsos (ed.), *Sustainable Mobility for Island Destinations*,
https://doi.org/10.1007/978-3-030-73715-3_3

Abbreviations

FTS	Flexible Transport Services
IT	Information Technology
ITS	Intelligent Transport Systems
MaaS	Mobility as a Service
PT	Public Transport
PTA	Public Transport Authority
PTO	Public Transport Operator
SUMA	Shared-Use Mobility Agency

1 Introduction

Mobility demand is becoming increasingly flexible in terms of users' segmentation and needs (both residents and tourists), transport mode, travel time. People are always on the move (work, education, leisure, health, etc). Connected mobility, involving Information Communication Technology (ICT) and Innovation, emerged considering mobility services as a seamless offer in terms of coordination of different modality and support service schemes, easy virtual and physical access, integrated payment and use of services irrespective of the mode and the operator. Users want to increase their flexibility in their choice of transport modes, but not being penalized in terms of accessibility, information, tariff schemes and costs. Some H2020 European projects have analysed these trends as CIPTEC (http://ciptec.eu) and MINDSET (http://www.mind-sets.eu/). On the one hand, these mobility trends could probably modify the future car use approach; on the other hand, they have already driven the growth of alternative ridesharing schemes.

On islands with rural areas and small towns, mobility choices of residents and tourists are limited compared to those in urban areas. Urban areas have been arranged so that people and all their needs are in proximity. In rural areas, destinations and activities are more diffuse, and the mobility solutions that work in urban areas are often unaffordable or unsuitable. In rural areas, it is hard to provide Public Transport (PT) services that are strictly tailored to the mobility needs of the various user groups (including the tourists) as demonstrated by the low use rate of PT services. The result is that small islands, small towns and rural areas become even higher car-dependent. The inevitable outcome is that cars are mostly used also for tourist travel. This is clearly incompatible with policies that seek to develop sustainable mobility for residents and tourists.

Last but not least, COVID-19 pandemic has added other trends affecting the demand: during the lockdown, mobility has been reduced to minimal activity and the prospects for recovery of user demand remain uncertain, with first data gathered in the Post-COVID period demonstrating a very slow growth. This can result in an excessive financial burden on traditionally lower demand services. While active modes such as walking or cycling might replace some of the trips by PT, there is the risk that many users shift to an unsustainable mode such as a private car (also for leisure and touristic trips). The Post-COVID protection measures (i.e. limited capacity on the vehicle, social distancing at the stops/terminals, etc.) add new constraints to the capability of "conventional" PT to be really effective in answering to mobility needs (becoming more differentiated in terms of time, places and target groups).

The COVID-19, therefore, has hugely impacted the transport sector with the risk of "new rise" of private car use also provoking new needs and other different options at the level of tourist destinations like the following ones:

- The municipality or the association of Municipalities needs to come back to a central role in Public Transport and collective services;
- Rethinking the space management policy in order to guarantee dedicated spaces to collective transport and active modes;
- Integration of occasional fleet in public transport service, ride/asset sharing schemes and personal mode (active, clean);
- Operation of a low-cost platform for control, coordination and reporting transport services offered in one or more areas;
- Operation of tools for monitoring the different services and providing specific services to the different user groups (residents and tourists);
- New partnership and business model for supporting the main mobility components (being the Agency one of these).

The above aspects push more the role of digitalization and the connected mobility in the policy and solutions for rural mobility, including also the tourist destinations.

The following Sect. 2 outlines the main mobility problems and needs concerning the role of innovation in transport services provision with respect to the other basic elements like mobility schemes, user information. Section 3 provides a picture of the emerging trends in digital solutions is provided, focusing the Mobility as a Service (MaaS) concept and the mobile applications. For the APP market specific considerations are developed in Sect. 4 on the different roles and typologies in order to face the current Mobility APP market, always more similar at a "Jungle". Moreover, the concept of Shared Use Mobility Agency (SUMA) is presented. Section 5 presents the SUMA (Gini, 2019) implementation in Elba island (Regione Toscana-Italy), within CIVITAS DESTINATIONS project in terms of needs analysis, design, implementation, organization and business model aspects. Finally, in Sect. 6 some key recommendations for Mobility Authority and Transport Operators are spotlighted.

2 Key Mobility Aspects

2.1 Mobility Demand and Transport Schemes

The role of PT services, as emerged by the initiatives and practices around Europe, independently of the operator (public, private or public–private partnerships), is the key tool for facing not only the mobility problems but also for supporting the citizens day life with different role in relation to the different areas (rural/peripheral areas, small-medium towns, metropolitan city, etc.) and transport demand requirements and segments.

PT, especially since the last decade, is changing its offer in terms of service schemes/options (e.g. demand responsive, ridesharing, targeted services, etc.), customizing them to the local needs and specific purpose/target clients and operating them as last-mile/feeder of the "conventional" services (serving primary axis/connections). Although public and private sector stakeholders need to cooperate in order to address challenges regarding regulation, insurance, business models and equity, shared mobility (asset and ridesharing), services can improve the accessibility of targeted areas, thus they need to be integrated into the whole mobility offer. It might be that Public Administrations and PT/Mobility Agencies take the lead in exploring the modalities of coordination/cooperation, playing a leading role in defining the framework and solutions.

From the demand side, a mobility offer restructured as above described, allows the engagement of potential customers who are available to use PT on an occasional basis by combining this choice with other possibilities; in a short time, the frequency of PT use can be increased. On the contrary, Public Authorities (Municipalities, Mobility and PT Agencies, etc.) usually consider the uptake of ridesharing services as a substituting solution for PT than as an integration element.

Ridesharing services cover a broad range of flexible and intermediate individual and collective transport modes: from dynamic ridesharing services to peer-to-peer transport schemes and new forms of "institutionalized hitchhiking" to well-established schemes like bike and car-sharing to collective taxi and carpooling.

From the service point of view, models promoted by ridesharing services under the "DRT or Flexible Transport services" concepts which have been already implemented from the 90ies by PT operators and authorities. Flexible Transport Services (FTS) could be adapted for meeting user needs, typically on a trip-by-trip basis with flexibility on three operational service parameters (routing, timing, vehicle used) to enhance service in response to demand. FTS include a wide range of services and schemes, such as general use, local and feeder services to trunk haul, replacement of low-frequency conventional services, service in low-demand periods, special services to specific users' groups, services in low-density rural areas, efficiencies in social mobility resources, niche urban markets, fuzzy lines between small buses and big taxis, etc. These services and schemes could be supported by the last decade improvement in digitalization and connecting devices.

2.2　Innovation and Digital Solutions

In collaboration with other initiatives as SMARTA project on shared mobility solutions in rural areas (https://ruralsharedmobility.eu/) project, based on the analysis of the different involved Pilot sites and the related mobility and technology solutions, a more in-depth discussion was dedicated to understand the correct relation among Innovation and Technology in Mobility and Transport services covering also the role and responsibilities of different stakeholders involved in rural areas and tourist destinations.

Very often, especially from the local public stakeholders, the approach to the innovation is based on considering the digitalization as the solution instead as a tool or enabler element for setting most suitable solutions answering to the identified mobility needs of the different user groups (tourists included). Moreover, if the identified solution is "digital", many times the choice falls on pure innovative advanced tool rarely applied in real operated service rather than on a stable and consolidated system. The innovation should be closely tuned to operation situation (not only from the PT and mobility services), characteristics and objectives of context areas considering the needs and requirements of the different citizen groups. Therefore the "Innovation", especially for mobility and transport services should be turned in different aspects of the service provision chain as the following ones: planning and engagement, business model and marketing product and service schemes, priority and impacts, etc. Innovation should be also considered based on the emerging scenario with a mobility demand more erratic and less systematic, increasing role of transportation network companies and asset sharing service operators, mutual interrelations among tourists and residents needs and requirements, decision-makers much influenced by digital solutions and social media.

As we analysed during DESTINATIONS implementation phase, some sites defined small innovative measures that allow facilitating the PT services and the use of active modalities, some other sites realized digital platforms for informing the users and for integrating the management and payment of the service. "Innovation" is not a unique definition and should be declined with respect to the objective of the city and the needs of the "users" groups. In the context of Mobility and Transport services for tourist destinations and rural areas, the Municipality can "innovate" "only" with the extension of the collective services and the improvement of the related quality through the implementation of shared mobility services in order to increase the flexibility of the overall mobility offer. The above "innovation" approach should be realized also with digital solutions, but surely the key role is played by the Bus, the first "type" of ridesharing service.

2.3   Lack of Integration of Mobility Services

The mobility offer even based on "different" type of innovation has not been yet fully able to answer to the problems and emerging trends described above.

Focusing on the integration of transport services/offer, one of the main aspects that emerged from the DESTINATIONS sites are the following:

- Lack of policy and methods for planning the standard transport services and shared services   on the basis of the different needs of the user groups and area characteristics;
- Poor coordination among different mobility operators and lack of multimodal options/solutions;
- Poor collaboration among public and private operators in order to offer complementary services (Demand Responsive Transport, sharing and active modes, …).

Therefore, as explored in Elba island, the setup of an integrated mobility offer requires the implementation of "tailored" flexible services customized to the needs of the target area and users; the flexible services should act as "feeder services" in relations to the main PT routes or main connections, the coordination and the exposition of the overall differentiated offer carried out by unique entity (Agency).

From the digitalization side, many relevant problems concern the network integration of the different solutions and systems (ticketing, control and monitoring, user info), although specific initiatives on data standardization and protocols have been defined at European and US levels mainly for the management static and dynamic service data. Moreover, the level of collaboration among the different actors/operators (public and private) involved in the transport service provision chain is still very low (or difficult to realize) impacting negatively on the level of intermodality and interoperability of the services. The main challenges are related to the holistic service management, business model, common tariff scheme, commercial agreement, revenue reporting, accountability, etc. All these aspects require a key role of mediation and coaching of the Local Authority that beside fixing the regulatory framework for the different services (transport fleet, mobility service, parking, etc.) should plan a coordinate mobility offer among the conflicting needs of the different operators.

In this context the main issues emerged during the DESTINATIONS implementation for improving an "integrated" and users-responsive mobility offer requires different interlocking dimensions:

- Acting on regulation, focusing on a specific PT and shared regulatory framework (which in various EU States is lacking);
- Developing new service concepts to meet evolving needs;
- Identifying the markets these will serve and the transportation function for each;
- Identifying, developing and deploying appropriate technical solutions to deliver the services;

- Developing the business case for the foreground and background services, defining appropriate commercial agreement among the involved operators in case of integrated payment;
- Defining an appropriate organizational framework for delivering transport services. In particular, with regard to the operation of ride and asset sharing services for specific areas and specific user needs;
- Establishing specific private–public-no-profit cooperation schemes for the operation of the services.

The above aspects have been faced in Elba with the implementation of the SUMA concept (Ambrosino & Gini, 2018) and are described in Sect. 5.

3   Last Decade Digital Solutions

In the last 5–10 years, the domain of Information and Communication Technologies (ICT) has experienced a quick and uninterrupted growth in the number and in the deployment of solutions. ICT recent trends range from remote software platforms (i.e. i-Cloud solutions for data retrieving, storing and management) to platforms/solutions for the integration and management of data generated by different systems and sources, from mobile applications and hand devices to the improvement of wireless communication networks, etc. As a result, in the mobility sector, Intelligent Transport Systems (ITS) deal with the deployment of new emerging solutions implementing new digital services for users/stakeholders, or offering consolidated services under innovative approach/accessing tools: from integrated platforms for the monitoring and management of transport resources and events to info mobility systems on SMS/mobile, apps/web portal, integrated payment systems (via SMS/credit cards/smart cards or electronic tickets, etc.) and support systems for service operation (fleet monitoring, traffic lights priority, etc.).

ITS have increasingly caught the attention of Public Agencies, Public Transport Operators, and mobility stakeholders: this interest has grown in parallel to the most recent ICT solution developments, driven by the perception that improvement of hardware/software solution performances and the enhancement of communication networks can immediately boost higher performances of ITS systems, allowing the introduction of new customer-facing services and, at the same time, decreasing the amount of human resources required (e.g. in on-street control and administration functions). Underpinning this is that all these factors produce significant economic savings for mobility stakeholders (in particular Operators) and, in general, a drastic cut of allocated public resources.

While this potential certainly exists (and is well-achieved in a variety of cases), it is necessary to pay a high attention to this point. The actual experience with ITS applications in a wide range of countries is that the expected performance and/or benefits are often not fully achieved, due that operational and organizational impacts produced by ITS and related costs are often underestimated.

A "key example" is the provision of multimodal real-time information on traffic and transport services operated in an urban or metropolitan area. These services are provided via web portals and mobile applications and are now rather frequent in many European and non-European towns. As already stressed, this trend is driven by the recent developments of solutions for the integration of data generated by different systems and tools (webservices, xml), the growing impact of standardization solutions for data geocoding (WSDL, WMS, WFS, etc.), data formatting and exchange (i.e. at European level: TRANSMODEL and SIRI for Public Transport data, DATEXII for traffic data, etc.) and the availability of technological solutions enabling an "always-on" approach to information access (smartphones and tablet, mobile app, web 2.0, etc.) to users. This kind of systems are able to access data from different sources (systems operated by different operators/stakeholders), aggregate and elaborate them for the provision of added value services: static and dynamic info on PT (timetables, location of stops, lines description, real-time arrival time of vehicles at bus stops, etc.), parking info (location, total number of lots, real-time availability of free lots, etc.), traffic events (road works, modification to the network, real-time identification of road sections with traffic congestion), flexible services (location of bike stations, real-time availability of free lots for bike, location of car-sharing and booking status of car-sharing vehicles), location of recharge points, multimodal journey planner and tickets payment.

Finally, besides the great opportunities of digitalization, some challenges need to be addressed, both technical, such as data collection issues, interoperability, scalability and information security, and non-technical, such as business models, usability, privacy issues and deployment.

3.1   Development of Mobile Technologies

Mobile technologies have an impact on both demand and offer side. Customers (demand side) want mobility services that meet quality standards as other services do. In response, stakeholders (offer side) adopt new approaches in their relationships with customers increasing services for mobile technologies and platforms, such as infomobility (e.g. multimodal apps etc.), payment (SMS, QR code, NFC etc.), customer care (e.g. use of social media, customer requests, customer satisfaction surveys) to engagement processes (e.g. online questionnaires). The rapid growth of MaaS initiatives aims to provide customers with an integrated offer that can be accessed based on their specific needs.

3.2   "Virtual" and/or "Remote" Platforms

The availability of "virtual" and/or "remote" platforms allows for re-design of flexible PT/mobility schemes in a way that has not been entirely embraced by Public

Authorities/Operators, instead of being primarily led by NTC providers. Innovative mobility services combine flexible schemes with traditional technologies, with new capabilities offered by "remote" platforms. Large benefits in optimizing resources and reduce cost impacts (both for investments and operation) could be achieved by exploring the possibilities for aggregating and networking different Operators in an integrated mobility offer through "virtual" platforms. New services can be more financially sustained if shared between operators or made accessible through "Service as a Solution" (SaaS)".

3.3 Open Data/Standardization

Open data and standardization foster the exchange of data among different actors (Mobility Authorities, Public Administrations, Mobility Operators, IT providers, etc.). Open data trend pushed the development of "added value" services (i.e. pre-trip/on trip information including static and dynamic info, interoperable payment solution such as credit card payment) for the customers and contributed to break down the market, giving the possibility to any IT provider to implement their applications in a competitive and fair way. Furthermore, both open data and standardization improve the fair and transparent flow of information between mobility stakeholder and external actors. From the perspective of Mobility stakeholders, the standardization of data format contributed to create aggregated data sources both at the level of a single organization (i.e. combining data generated by ITS implemented by different IT provider) and at the level of the served area (i.e. combining data produced by different Operators). Aggregated data sources are the pillar for the introduction of "data mining" tools to leverage the "business-oriented" approach of mobility sector: service planning optimization, tuning of operational procedures efficiency, assessment of service performance as a baseline for the definition of improvement targets, etc.

3.4 MaaS (Mobility as a Service)

MaaS (Mobility as a Service) concept emerged with aim to solve the current fragmentation in ticketing, info, marketing, accessibility and cooperation among different mobility services and operators. MaaS can be defined as the transition from a dominant car-ownership model of mobility, to one of paying for mobility "on account" using more sustainable modes of transport as an alternative to private car use (Hietanen, 2014). With MaaS, individual transport needs (usually satisfied by car), are instead met by a range of services such as car leasing, car clubs, carpooling, community transport, bike and taxi services in combination with "traditional" PT services. MaaS concept in any case has been developed mainly on urban/suburban areas.

E-ticketing, AVM and User information systems are the base systems for any MaaS concept implementation. In Europe, it is common that, these systems are designed, contracted and operated under the responsibility of PT Operators or Agencies but not always they are planned under a "city approach". These systems are often designed individually and their integration must be carried out on a "bottom-up" approach.

The role of PTOs is fundamental in MaaS as PT services are the backbone in terms of data provision, responsible for system operation and management and for assuring the quality and reliability of services on the road. None high-quality MaaS service will be possible without an efficient and integrated PT offer and without the certification of the data quality provided by the base-level ITS.

The operation of a wide range of ITS dealing with MaaS must guarantee high system performances and strong reliable of services and data. To achieve these objectives, the involved systems must be operated with a well consolidated organizational structure and proper operational procedures: suitable data certification procedures are required and high-quality data must be assured by each Operator based on contract terms and related rules which should be issued by the coordinator of MaaS initiative (EMTA 2019).

In any case, the experiences on MaaS realized until now are related to urban context like large or medium-size towns with PT operated under the direct responsibility of Public Authority. The MaaS concept in rural area where the PT offer is very low and based on different mobility schemes and operators is still unexplored for the lack of a relevant demand and for the fragmentation of existing mobility demand.

At this stage after almost 10 years since the launch of the MaaS concept some considerations can be summarized as the following ones:

- There is not a unique MaaS definition, each stakeholder or market provider has a specific definition and implementation of MaaS concept;
- MaaS is a process to be realized gradually depending on the context mobility objectives and the existing systems/services;
- The risk is to use new "acronym" to hide unsolved challenges: organization, operation, service quality, commercial deal aspects, contract management and system reliability monitoring;
- It is vital an effective cooperation among the different Authorities/Operators responsible for the different systems and services under the indications of the ITS EU directive;
- The role of ICT/system providers is not secondary and they should be involved in the engagement and cooperation process.

The above considerations bring also to highlight some issues or questions that should be faced for consolidating the MaaS concept like the following ones among the others: *Which regulation framework* for the service provider? *How to guarantee the quality* of involved services? *How to guarantee quality* of system data production? *How to appoint specific systems* in MaaS in terms of performance indicators? *Which role for the PP Partnerships* approach and related business model?

A tentative to facing some of the previous aspects and question has been carried out for the rural and tourist areas with the implementation of the Shared Use Mobility Agency, described in the section below and detailed in Sect. 5.

3.5 Concept of Shared Use Mobility Agency

The SUMA concept aims to reconcile and enhance two parallel axes of mobility services (collective transport and asset/ridesharing services) by common planning, coordination, management and integration of the different innovative mobility solutions.

Both public (collective) and private levels interact with each other: major transport axes and corridors on one hand, and flexible/ride/asset sharing services on the other. Fostering the interaction between public and private mobility through various connected mobility schemes (parking, Park and Ride, interchange facilities with shared vehicles' schemes, integrated payment, etc.) is pivotal for improving mobility as a whole.

Effective integration between collective and shared mobility is achieved by working on three interrelated levels (collective transport, personal mobility and connected systems). SUMA offers integrated access to several "on-demand and shared" individual and collective services, by coordinated management of the various actors and services through an ad-hoc organization framework, technology-enabled services and soft measures.

SUMA also acts as a base for any MaaS initiative tailored to mobility demand and the territorial context and the needs and requirements of citizen groups (Ambrosino & Gini, 2016).

3.6 Key Role of Mobile APP

Despite facing different problems, objectives and approaches, all the above scenarios and trends objectives and approach give a central role to Mobile APPs and this role is differentiated for each digital solution.

In this context, the emerging APPs in mobility and transport services are linked to the most recent ICT developments described above and they have the following objectives:

- to improve the overall PT marketing strategy, not only providing users with information on the services operated, but also establishing an innovative channel of communication and interaction with them (functionalities feedbacks);
- to collect service data for acting on the service operation and/or to evaluate the service planning and program compared to the real service behaviour;

- to integrate information and services already available/provided in a number of separate channels/modalities, in order to create a unique "contact point" with clients and to optimize and enhance the technological investment already made;
- to build on the standardization and the modularity of the technological solutions, in order to allow the future extension to other platforms (web portal), other mobility services (also provided by other Mobility Operators) and to "Open data" publication.

In the last decade, many different APPs have been introduced in the market both by the transport and mobility operators and new emerging Transport Networking Companies (from UBER to MOOVIT). These APPs have different objectives and roles from provision of innovative infomobility for PT and mobility services to the integration of info and the provision of multimodal added value services, the identification of the mobility needs and the service coordination.

Usually, the transport operators are highly committed in the introduction of the infomobility to increase the quality of transport services provided to users and citizens, to improve the company competitiveness and to optimize/unify internal procedures for delivery of the services. As a result of this approach, many Mobility Agencies or Transport Operators manage a wide range of innovative services to provide information to users: info panels at bus stops and terminals, on-board monitors for next stops visualization and infotainment, PT trip planner and info on timetable, line diversions and relevant news on the web. In this context, the APP is identified, by the PTO/PTA as a new product for the enforcement of commercial initiatives addressed to customers and potential users accustomed to smartphone/tablets and mobile services (H2020 2018).

The APP implementation usually is based on a platform able to access a wide range of base data stored/generated by the other company systems/tools (database, applications, web services/file transfer), to update, integrate and elaborate such data in a common standardized layer (middleware) and to provide added value services on mobile devices (Android and iOS platforms) and web portal (future extension). The setup of the standardized middleware will be an added value in case of future extensions of the platform to other mobility services and the provision of "open data" to third-party B2C services developers. The APP is designed to manage data for the various service areas operated by the same PTO company and to be multi-language (Italian/English in the version already published but compatible with the extensions to other languages).

Also, in DESTINATIONS different APPs have been developed. Based on an analysis and classification of the APP Market, carried out in collaboration with SMARTA project, criteria have been defined for choosing the most suitable APP in the existing APPs "jungle".

4 APPs Jungle

This section details the key issues to be addressed for the implementation of APPs in the mobility sector. Initially, a classification of the APPs is provided in order to set the main scenarios (Gini, 2019). In fact, when talking about APP, the term "service" indicates a function or a set of functions allowing the users to receive information on a specific mobility (or other sectors) services (i.e. real-time information on bus arrival, journey planner, etc.) or to make an operation (i.e. to buy the ticket). The use of the term "services" in such a context should not generate confusion with the mobility services which are operated on the network (e.g. Public Transport (PT) or demand responsive) producing information (data) for some APP functions (i.e. bus arrival delay).

From the technical point of view, the APPs operated in the real environment are usually a combination of two (or more) different categories presented here; nevertheless, the suggested approach is useful at the beginning of the design phase to identify the required functionalities and then to analyse the required data. The following three main typologies/categories can be identified:

- Class-A ("CA") APPs exposing information services (i.e. PT timetable, PT real-time information). For this category, the mobility/transport services (producing the data) are operated by a Public Transport/Mobility Operator in one or more areas (or Town, Province, Metropolitan areas);
- Class-B ("CB") APPs exposing information services and allowing a wider set of end-users' services (e.g. tickets payment, journey planning, etc.). The services are operated by a Public Transport/Mobility operator in one or more areas (as in "CA");
- Class-C ("CC") APPs integrating data/services from/for different Mobility Operators (in one or more areas) and/or extending the end-users' services to innovative services such as crowdsourcing, management of relations with the customer, demand behaviour triggering (gamification), demand aggregation (ridesharing), feedbacks from the users (i.e. service quality assessment), O/D matrix estimation, etc.

4.1 Exposing Information Services to the End-Users (CA)

In order to easily describe this APP category, it is possible to focus "CA" APPs on those APPs whose services relate to a single Operator (the service can refer to one or more covered areas).

This type of APPs provides information services to the end-users, for example:

- Info on the scheduled service (static data such as timetable, graphical representation of the lines on the map, schematic representation of the line, list/map representation of the bus stops-pick up/drop off points, etc.);

- Single-mode/service (PT) journey planner (eventually integrated with walking info);
- Info on the operated service (dynamic data such as the position of the bus on the line, real-time arrivals of the buses at the next stops, modifications of the operated service compared to the scheduling, line diversions, events on the network, etc.);
- General information for accessing and use the service (static data) such as a description of fare and tickets, list/map representation of the selling points, access rules to access the vehicles, procedures for reclaims in case of irregular service events, etc.;
- News links (to company or institutional websites), etc. This could include "external links" to other APPs or websites providing "CB"-type services (i.e. tickets payment).

MOOVIT[1] (Fig. 1) can be considered an example of this category, despite different Mobility Operators (Public Transport, asset sharing, etc.) are included and the APP allows also the collection of the notification/feedbacks from the end-users.

4.2 Exposing Information and Providing Access to a Wider Set of Services (CB)

As the category "CA", the "CB" category APPs still refers to the services provided by an Operator (in one or more covered areas). The "CB" category APP includes the type "CA" and provides the end-users also with access to a wider set of services, including:

- Tickets purchase through different modalities (SMS, QR code, bank card, NFC, different "wallet functions" available on the market such as GooglePay, SAMSUNG Pay, Apple Pay, etc.);
- Recharge of contactless smart cards or e-purse (directly via NFC or after the smart card is put near a validator or an Automatic Vending Machine);
- Booking/seat reservation.

TreviMove[2] (Fig. 3), the APP operated by Mobilità di Marca (the mobility Operator in the Shire of Treviso, Veneto Region) and Tiemme Mobile[3] (Fig. 2), the APP operated by Tiemme (the Public Transport Operator in the Southern and Eastern part of Tuscany, one of the largest Italian PT operator) are examples of the "CB" category. Nevertheless, in the case of TreviMove, the services provided by the APP are extended to parking services which are operated directly by MoM in the urban area of Treviso.

---

[1] https://moovit.com/.

[2] https://mobilitadimarca.it/p/trevimove-app.

[3] https://www.tiemmespa.it/.

Fig. 1 MOOVIT Graphical representation of bus stops and line with real-time arrivals

4.3 Integrating Operators Services, Innovative Functionalities (CC)

The "CC" category APP can network different mobility services (but also no-mobility services, as the tourist ones) and provide a broader set of services compared to the "CB" category.

With regard to the networking of different Operators and services, "CC" category APP provides the following "services" as examples:

- Integration of the static/dynamic information related to the services provided by different Public Transport Operators (PTO) in the same area(s), i.e. different bus

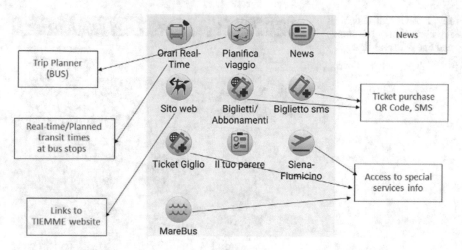

Fig. 2 TIEMME and key services

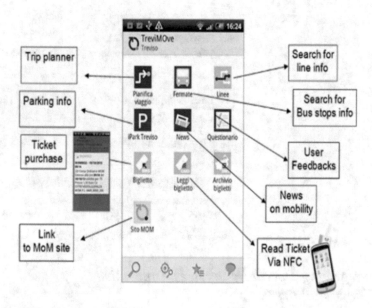

Fig. 3 TreviMOve key services/functions

Operators, conventional + flexible services + niche/dedicated services, bus + train, etc.;

- Integration of the static/dynamic information related to the services provided by Mobility Operators in the same areas (i.e. parking areas (location, number of available lots, occupancy, etc.), related to Park&Ride services, bike-sharing as PT feeder, bike stations, rental (bike) operators, access restrictions to city centres, etc.);

- Integration of static/dynamic services related to other mobility services (integrated payment – MaaS, booking of services package, etc.);
- Multimodal journey planner (private + PT, eventually integrated with info for cycle lanes).

As extension of the Mobility/Public Transport APPs, "CC" category can encompass info more properly related to tourism experience and overall accessibility/travel needs in the target area (i.e. asset sharing/rental operators including scooter or cars, hotel-bike service, list of tourist destinations, etc.) and the services can be extended in parallel (services for tourism accessibility, integration between mobility and tourism services, etc.).

Concerning the integration of innovative services, some examples are provided in the following:

- Interaction with on-board devices (eventually installed) for visually impaired customers;
- Collection of data by crowdsourcing initiatives (i.e. estimation of Public Transport O/D demand matrix through automatic detection of APP connections or registration of trip diary);
- Collection of notifications and feedbacks from the customers (online surveys for needs assessment, online questionnaires on the quality of service provided, suggestions for service improvement, etc.);
- Gamification/management of incentives (i.e. discounts for commercial products) to promote the use of sustainable transport modes and eco-friendly travel behaviour;
- Aggregation of the demand between customers and management of ridesharing service (when the end-user acts as transport provider itself).

To get some examples, it is possible to look at:

- the ROGER APP[4] (Fig. 4) launched by Regione Emilia Romagna (as an example of the integration of static and dynamic data to provide multimodal and multi-operator services on mobility (PT + parking) at regional level);
- the APP developed by the Municipality of Almada (Fig. 4) within INTERREG MED MOTIVATE project, as an example of trip diary registration and collection of end-users feedbacks;
- the Elba Sharing APP[5] introduced by the Municipality of Portoferraio within CIVITAS DESTINATIONS project (see Sect. 5);
- Stadtnavi[6] (Fig. 4) provided by the city of Herrenberg for route planning in Herrenberg region. The service covers PT, walking, cycling, private car use and carpooling.

---

[4] https://rogerapp.it/.

[5] https://play.google.com/store/apps/details?id=net.pluservice.elbasharing&hl=en.

[6] https://stadtnavi.de/.

Fig. 4 ROGER, Almada and Stadnavi APP

As complementary consideration, it is possible to state that the choice of the APP category is also related to the specific context, end-users target and background conditions. To give some examples, the selection of the functionalities to be included can be driven by the relevance of the various end-users segments (in particular some functionalities may be targeted to specific segments such as elderly or people with reduced mobility, others may be targeted to receive feedback from passengers on the quality of the services f), the opportunity to add the ticket purchase function depends on how easy it is to buy a ticket with other selling channels, etc.

5 The Elba Shared Mobility Agency

The SUMA concept aims to reconcile and enhance the two parallel axes of urban mobility (collective transport and flexible/ridesharing services) by testing and demonstrating different innovative mobility solutions to be integrated under its "umbrella" (Nelson & Ambrosino, 2015). The concept can be tailored to different contexts selecting the functionalities and adapting them to the needs (Fig. 5).

One of the main measures of DESTINATIONS project was the development of the SUMA in Elba as centre for managing and making accessible several "on-demand and shared" individual and collective transport services, coordinating various actors and services supported by an advanced IT platform.

A large part of the MaaS initiatives in Europe has been implemented in urban areas (White Paper on Mobility as a Service (MaaS) and initially pushed by commercial operators (such as MaaS Global in Finland). When the implementation phase moved from early adopters to a more mature status, it has been found that Public Authorities and Mobility Operators need to take the leadership of MaaS initiative as they are the main data provider (a perspective on MaaS from Europe's metropolitan transport authorities, EMTA). The implementation of SUMA in Elba aims to adapt MaaS concept to rural areas (which have quite different needs compared to the urban area in

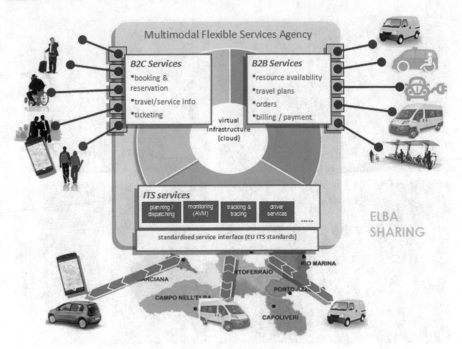

Fig. 5 Concept of Shared Mobility Agency (SUMA)

terms of flexibility of demand, type of mobility offer and mobility solutions). In large urban area metro/tram/light rail are the key transport modes, PT network is based on main axis and demand is more related to commuting habits. Moving to medium-small sized cities and rural area bus is the main transport mode and sharing (asset/ride) services are complementary rather than competitor of bus services. Demand is more dispersed in terms of destinations, time and purpose for travelling. Furthermore, MaaS implementation in Elba is strongly led by the Municipalities (Portoferraio) supported by the involved Public Transport Operators (Gini & Ambrosino, 2019).

5.1 Reference Context and Users' Needs Analysis

Elba Island is located 10 km away from the coast and is the main island of the Tuscan Archipelago (Fig. 6). Consisting of seven municipalities, Elba has numerous towns and small settlements spread across the island with predominantly rural areas in between them. The island's PT services operate mainly along its key transport arteries.

Elba is one of the main tourist destinations in Europe, not only for Italians but also for foreign travellers coming from Central Europe (Austria, Belgium, France, Germany, Netherlands and Switzerland) and looking for blue sea, beautiful beaches

Fig. 6 Elba context and tourist attraction

and outdoor experiences. The annual number of tourists (pre COVID situation) is 450,000/year with a daily presence of about 30,000 people/day bringing a 30% increase in transport demand, alongside a drastic variability for destinations (beyond the main transport network) and time (low-demand hours beyond peak hours).

Currently, the island's PT system cannot meet these needs and this situation is further exacerbated by the fragmented nature of Elba's mobility offer (including rental operators): services are uncoordinated and information spread across various sources affecting the way a tourist can access them. A coordinated mobility offer, including collective and private modes, is an added value in Elba, sustainability and overall accessibility and could also bring new job opportunities.

The problems that affected the mobility offer in Elba are (Fig. 7):

- High seasonal mobility demand (early morning to late night);
- Dispersed origins, few concentrated destinations (beaches, discos/bars, camping, museums and archaeological sites, etc.);
- Low use of PT services (only 14% of overall trips); private cars and motorbikes are largely used but, in some cases, are not the most suitable transport mode due to parking scarcity and poor accessibility of a wide range of destinations;
- The offer of rentals (bike, scooter, car, boat) is increasingly high, but it fails for coordination.

SUMA provided answers to the identified local needs:

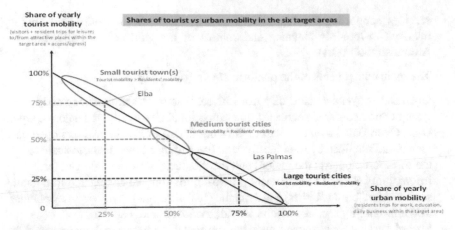

Fig. 7 Share of tourists against residents for the six destinations cities

- A coordinated mobility offer "brokered" by a unique Agency will be an added value in Elba attractiveness, sustainability and overall accessibility;
- The design of the Agency is based on the networking of the local vehicles (i.e. bike/scooter) operators on the island in order to optimize the available resources and to create a suitable offer to be integrated with PT;
- The user can act as mobility service contributing to solve the conflicts between the PT services and the individualized mobility.

The design of SUMA (carried out by MemEx) consists of three different levels:

- The technological platform supporting the Agency operation;
- The organizational and operational level for daily management;
- The business model for long-term sustainability.

The following sections, detail how each of these areas has been worked out.

5.2 The Specifications of the Agency Platform

The enabling technological infrastructure for SUMA is based on the emerging paradigm of the Internet of Services or Service Orientated Approach, providing several core facilities including:

- services for transport users (Business-to-Consumer B2C) enabling access to information, search for transport options, travel planning;
- services for the coordination of different ridesharing and mobility schemes and interaction with relevant operators (Business-to-Business B2B);

- services supporting the interactions among different authorities and entities involved in mobility planning and control of transport services (Business-to-Administration B2A).

The main components of the platform are shown in Fig. 8.

- Open Data Layer accessing data from different sources, standardizing and aggregating them in order to enable the other functionalities and the provision of Open Data. Open Data Layer accesses to different data sources: the regional open data web portal provided by Regione Toscana (providing a scheduled timetable for all the transport modes in the region) and local databases (for rental operators);
- Infomobility Service Management providing multimodal aggregated info (static such as timetable and tariffs) and dynamic (real-time) including a journey planner. The infomobility services will be provided on APP and web portal;
- Shared Service Management allowing the matching between demand and offer for sharing trips;
- Operators networking enabling a centralized "window" for the offer of rental services.

These functionalities represent the first core of SUMA which can evolve in the future up to a "full" MaaS concept for rural/peripheral areas (including the integrated payment).

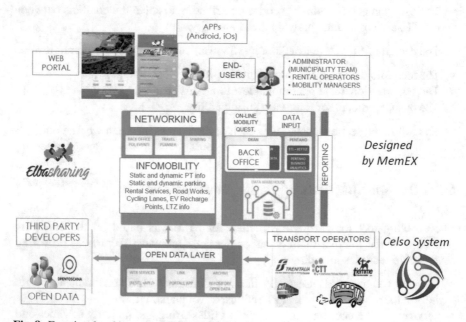

Fig. 8 Functional architecture of Shared Mobility Agency for Elba

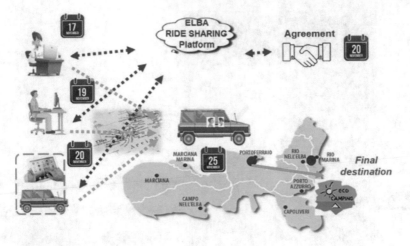

Fig. 9 Trip sharing planning

5.3 Use Cases

Users registration is required to use the functionalities for the management of ridesharing services. The registered data are stored and handled in compliance with the national privacy regulation and EU requirements (General Data Protection Regulation, 2016/679). The management search engine function shows to the user (Fig. 10) the options to combine ridesharing offers and requests with the use of PT and the offer of rental operators. Ridesharing possibilities are numerous, including sharing a trip when a user is on the ferry; planning shared trips in advance, and sharing a taxi trip Fig. 9). These scenarios can also be managed through a notice board on which passengers and drivers can publish their trip share requests and offers.

Features such as journey tracking and user evaluation of vehicle drivers will also be included to improve passenger safety. Key use cases will be detailed in the presentation.

5.4 The Agency Implementation and Business Model

The call for tender to subcontract the platform has been prepared in October–November 2017 and launched in December. The ITS platform has been purchased as a supply including technical services supporting the implementation, training of operators and a three-year warranty. The bidder proposals (from three national IT providers) were received in February 2018 and the winning was awarded in June 2018. The Contract between the Municipality of Portoferraio and the awarded bidder was signed in September 2018. A first prototype of the Elba SUMA was released in 2019 including a first core of scenarios for the management of sharing services,

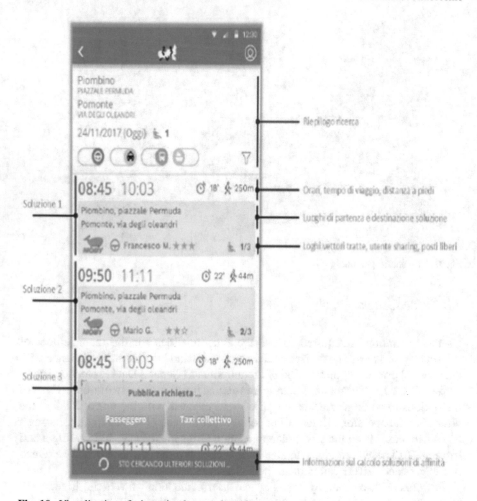

Fig. 10 Visualization of trip option by search engine

infomobility and networking. Two further versions of the IT platform were released during the 2020 winter and the final version with full functionalities is available from June 2020.

In the meantime, the operational procedures for the management of the Agency have been defined:

- Monitoring of data gathering processes from connected sources;
- Updating/management of data repository for SUMP (from "paper" surveys);
- Updating/management of static contents;
- Engagement of the networked operators of vehicle sharing services.

The following cost categories have been considered: internal staff (two operators are estimated to operate the Agency), software maintenance, utilities and promotion.

Moreover, the costs of specific technical support to Portoferraio Municipality should be also taken into account.

Two main segments of target customers will use SUMA: tourists and residents. Other stakeholders involved in the use of the Agency are rental operators, commercial operators (i.e. discos, restaurant, shops, etc.), third-party developers accessing the Open Data for their applications, Public Administrations, consultancy companies and professionals accessing data/survey stored in the Open Data Layer.

The economic viability of the Agency, after the end of the CIVITAS DESTINA-TIONS project, could be guaranteed by the external revenues deriving from these alternatives:

- an annual fee (to be established based on the performance of the first two years of operation), to be requested at the time of registration to users (both users and commercial activities/operators). Assuming that 5% of these are registered, equal to about 22,000 people, a minimum contribution of € 3 per registration could be reasonable, for a total of about 60,000 € annually;
- an annual contribution from renting operators (bikes, cars, scooters, boats, etc.) that are included in the network through the platform (with the possibility of dynamic interaction);
- a contribution from interested commercial operators (e.g. discos, restaurants, supermarkets, etc.) to be indicated in the platform as primary points of departure or arrival of "shared trips", in order to increase their attractiveness and safety level;
- possible contributions from local administrations or consultancy companies interested in the documentation and data on mobility and transport accessible through the Open Data Layer of the technological platform.

5.5 First Results

The final version of Agency was launched, in June 2020.

Unfortunately, the COVID-19 affected the possibility to carry out an extended and deeper test of the platform and Agency functionalities both from the specific residents' groups (especially students and village representatives) and by the different tourists available to leave the car at the hotel and to use the different mobility options integrated by the PT services offered by the SUMA.

In any case, the scheduled tests have been envisaged with a two steps procedure:

- A first session carried out by a selected group of people (including some tourists) in order to identify main problems (if any) which can prevent the use among a wider group of testers;
- A second session open to a wider group of testers recruited through specific engagement actions among residents: students, tourist operators, etc.

The first step, despite the difficulty of COVID-19, was carried out to make the first updates for the launch of the second version of the platform.

The promotion of the Agency and the related APP is scheduled with the second test steps focused on the improvement of user experience.

The launch of this version of the platform will allow checking which are the impacts of SUMA in terms of users and then the overall success of the initiative. The initial estimation, also based on the analysis of this (COVID-19) Summer 20 is to get at least 5% of the tourists visiting Elba (equal to 22.000 people). A minor part of users should also come from the visitors (to be added to the previous figure).

Once SUMA is entirely in operation, the allocation of resources estimated in the design phase will be verified and tuned together with the responsibilities and working procedure for daily management.

The feedbacks coming from the Agency operation (in terms of users, operational costs and level of engagement of stakeholders, i.e. the rental operators) will allow to validate the business model defined in the design phase and to adjust the figures if necessary.

6 Conclusions and Recommendations

The above sections introduced the role of innovation and digital solutions in touristic destinations and rural areas characterized by a low and variable mobility demand. For each set of digital solutions, some common practice recommendations have already been provided.

In this section, some other key recommendations are provided to Transport and Mobility Operators and/or Public Authorities (Municipalities, Agencies, etc.) aiming to implement digital solutions especially the APP for the infomobility and service management.

As already stated, the market of APPs is too much varied; therefore, the choice of a specific app should depend from one side on the different targets and required functions and from the other side on the different data sources to be linked.

For this, the concept of APPs "Jungle" and the categorization of the APPs was introduced, in order to support user needs analysis, to identify the key functions and screening the same "jungle". Therefore, the selection of APP typology and functions should be based on some criteria, among the others, related to objectives of the service implementation, target end-users, main needs in the accessibility to PT, institutional context and cooperation levels with the different stakeholders or with the other digital solution initiatives. These criteria should be analysed taking into account also the background conditions (i.e. available data, data collection tools, existing IT systems acting as data provider, system responsibilities etc.).

For the relevant issue related to "how" to get data the interrelation between open data and centralized data source should be explored in detail. In particular it should be clarified if the required data are already published by the owner or if required data coming from different services/operators have already been stored in a common data source at different levels (local, regional) to be interfaced. Another macro aspect that should be taken into account for defining the suitable digital solution based on APP is to require close cooperation among the involved providers/stakeholders in case the needed data sources are managed by different stakeholders or operators: it is essential that one stakeholder (i.e. the Public Administration, at city level) coordinates the different actors involved. The availability for cooperation by the technical provider of legacy systems and background applications is also required.

Special attention must be paid to the design of attractive graphical layout of APP/web portal; it is particularly important to consider the time required for an agreement on this aspect when a relevant number of local stakeholders are involved.

Even if the APP is well designed and duly implemented, its diffusion, use and acceptance can hardly be achieved if a specific promotion is lacking; therefore, it is strongly recommended to define and develop a relevant marketing strategy and actions.

Moreover, the complexity should not be underestimated, in particular:

- the evolution of APP layout and functionalities over time in terms of: ongoing—needs, data integration, contracting, acceptance, etc.;
- the quality of data and reliability of the functionalities in terms of performance indicators, testing procedures, data maintenance, etc.;
- smooth operation in terms of monitoring procedure for data accessibility, the responsiveness of operational procedure, back-office activity and operation/interfacing procedures.

Last but not least, the contract requires specific attention, especially in the case of digital solution procurement and implementation. The main aspects to be paid attention to are the following, among the others:

- modularity and future scaling-up of the solution not only from the technical solutions adopted, but also from the contractual point of view, thus avoiding extra-costs (i.e. validity of software license without time restrictions and without limitations of data sources interfaced);
- continuous support by the IT provider during implementation (interfacing the legacy providers, on-site activities for integration, installation and testing, maintenance);
- implementation milestones should be defined in order to manage step-by-step payments on the basis of the results of the testing procedure (functional and performances verification) to be carried out during the realisation phase;
- final payment and warranty should be released at the final acceptance of the whole system.

Finally, besides the above recommendations a "common sense" indication related to complex digital solutions implementation, like SUMA, is to keep the realisation approach simple.

References

Ambrosino, G., & Gini, S. (2016). Mobility as a service: The Italian Style. In: *ITS passenger transportation systems and services sponsored by ITS America and American public transportation association, 95th TRB*. Washington.

Ambrosino, G., & Gini, S. (2018). Implementing a new business model for flexible mobility. Intelligent Transport on-line Magazine. Retrieved from https://www.intelligenttransport.com/transport-articles/68818/the-flexible-mobility-concept/.

European Association of Metropolitan Transport Authority (EMTA). (2019). A perspective on MaaS from Europe's metropolitan transport authorities. Retrieved from https://www.emta.com/spip.php?article1319&lang=en.

Gini, S., & Amborsino, G. (2019). "MaaS approach in Elba: The shared use mobility platform" Booklet. Retrieved from http://www.memexitaly.it/wp-content/uploads/2019/04/CivitasDestinations_v2_April19.pdf.

Gini, S. (2019). Shared use mobility agency: Concept and implementation in Elba. In: *Destinations Technical Workshops, Elba Project Meeting*, Portoferraio. Retrieved April 10, 2019.

H2020 Urban Project. (2018). White paper on mobility as a service (MaaS). Retrieved from https://jpi-urbaneurope.eu/app/uploads/2018/09/JPI-UE_MaaS_white_paper2018.pdf.

Hietanen, S. (2014). 'Mobility as a Service'—The new transport model? ITS & Transport Management Eurotransport. Retrieved from http://www.itsineurope.com/its10/media/press_clippings/ITS%20Supp_et214.pdf

Nelson, J., & Ambrosino, G. (2015). Shared use mobility services in a mid-size historic town. In: *Annual Polis Conference*. Retrieved November 2015, Bruxelles.

Co-creating Sustainable Urban Mobility Solutions for Guests and Citizens: The Experience in the Elba Island

Carlo Sessa

Abstract This chapter illustrates a co-creation approach engaging stakeholders and citizens to develop a shared agenda of sustainable transport solutions, responding to travel needs which are qualitatively and quantitatively different for residents and tourists. The chapter is articulated in three sections. The first section illustrates the co-creation approach and the topics addressed in all six DESTINATIONS demonstration sites where this approach has been used to a greater or lesser extent. The second section describes more extensively the application of the co-creation approach in the Elba island case, illustrating the process and its main outcomes, including the definition of a strategic agenda of goals and actions included in a draft Sustainable Mobility Plan for the whole Elba island, and a hypothesis of new bus services operating in the high season co-created with the engagement of managers from the regional Public Transport Authority (Osservatorio della Mobilità—Provincia di Livorno) and from the Transport Operator (CTT-Nord). The third section discusses some lessons that can be drawn from the successes and failures of the co-creation approach in the Elba case. The application has demonstrated the feasibility and suitability of the co-creation process to effectively engage stakeholders and groups of visionary citizens (the "change agents") in the formulation of sustainable mobility plans and agendas. On the other hand, follow-up was lacking—no formal procedures to implement decisions based on the shared agenda were initiated—and the concluding section of the chapter discusses some reasons for this failure, and possible remedies for better anchoring the co-creation approach in future policymaking

Keywords Co-creation · Seasonal mobility · Stakeholders engagement · Scenarios

Abbreviations

LL   Living lab
PT   Public transport

C. Sessa (✉)
ISINNOVA, Rome, Italy
e-mail: csessa@isinnova.org

© The Author(s) 2022
T. Tsoutsos (ed.), *Sustainable Mobility for Island Destinations*,
https://doi.org/10.1007/978-3-030-73715-3_4

SMI   Small-medium island
SUMP   Sustainable urban mobility plan

1   Introduction: Stakeholders Engagement to Co-create Sustainable Mobility Solutions

The amenities of small to medium island (SMI) destinations attract a population of tourists that share living in the islands with the resident population. Depending on geographical and accessibility circumstances, visitors can enjoy staying on an island for a day, a weekend, one or more weeks, or even for several months. As a result, the varying concentrations of tourists and the different duration of their stay entail different mobility challenges in different periods of the year. Residents and tourists use the same transport infrastructure, usually limited to a road network, as the critical mass for building a rail network is hardly achieved in SMIs. Moreover, SMIs usually can be accessed only by ferry at one or a couple of main ports, or with private boats. Due to the lack of suitable and affordable alternatives to private car use, a large proportion of visitors, therefore face an additional cost as they embark on their car even for short visits.

The Elba island in Italy—located only 1 h of ferry journey away from the Piombino port on the Tuscany mainland—is an emblematic example of SMI with serious mobility problems, which shape and constrain the economic development and quality of life on the island for the residents and tourists alike.

In the following sections, we illustrate a participatory process engaging groups of citizens and relevant stakeholders to set an agenda of priorities to improve future mobility to, from, and within the island, and the first steps undertaken in the context of the DESTINATIONS demonstration project to single out a fully fledged ELBA Sustainable Urban Mobility Plan (SUMP).

1.1   The DESTINATIONS Challenge: Sustainable Transport Solutions to Meet Residents and Tourists Demand

In any touristic destination, mobility is a key component of the daily experience of residents and tourists alike. The purposes and needs of mobility are, however, different for these two categories, reflecting seasonality, the varying daily demand patterns, and the typologies of mobility services that are needed to satisfy these demand patterns.

The common challenge for all six DESTINATIONS sites[1] is how to respond to travel needs which are qualitatively and quantitatively different for residents and

---

[1] Madeira, Limassol, Las Palmas, La Valletta/Malta, Rethymno, Elba.

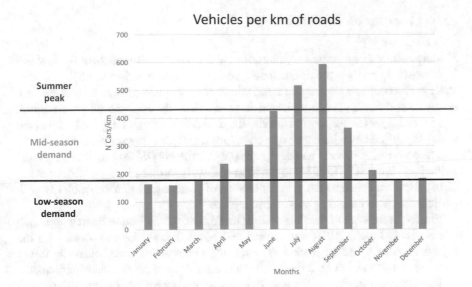

Fig. 1 Seasonality of transport demand in the Elba island

tourists. This challenge is exacerbated in small to medium size tourist destinations such as the Elba island—the smaller of the six demonstration sites, with a population of about 30,000 residents that grows at least fivefold in the high season.

As shown in Fig. 1 below related to the Elba island "cars carrying capacity", the seasonality of transport demand is really a wicked problem for small-medium tourist destinations, as the whole transport system supply (road network, public transport services, parking space, etc.), which is designed for the resident population in the low season (winter), finds itself severely under stress already in the mid-season (May, June, September), and dramatically so in the high season (July, August).

The proxy used to measure the "cars carrying capacity" is the number of vehicles per km of road infrastructure. The latter is dimensioned for a rural territory—the whole island hosting overwinter about 30,000 residents mostly concentrated in Portoferraio (about 13,000 inhabitants) and for the rest distributed in 6 other small municipalities—with a pressure of about 200 vehicles per km in the winter season. This pressure, however, grows to a range between 200 and 400 vehicles in the mid-season, and rises to something between 500 and 600 vehicles on an average day in July and August, with the inflow of private cars embarked in Piombino.

Moving around the island in the high season is difficult, congestion on the roads and at the parking spots available at the beaches and in the small towns is high, which severely constrains the freedom of mobility of both residents and tourists. In July and August, the island experiences a level of demand equivalent to a city of 150,000 inhabitants, but the transport infrastructure (the road network) is unchanged, and the supply of transport services (bus lines, flexible services) is direly insufficient.

1.2   Co-creation Methodology

Finding sustainable transport solutions for the different requirements of the low and high season is really a major challenge, which was addressed in the Elba demonstration case engaging local and regional stakeholders and groups of citizens in a co-creation process. The Elba island is an emblematic example of a polycentric region, and it was necessary to consider the different needs and requirements of the seven municipalities within a unique co-creation and planning process.

The co-creation process was designed applying the Poly-SUMP methodology and guidelines (Poly-SUMP, 2014) to the Elba reality. It further relied on (Kocak et al., 2014). for what concerns the peculiar aspects of planning sustainable mobility with participatory approaches in polycentric regions.

Finally, at the core of the Poly-SUMP methodology, the Future Search approach (Weisbord & Janoff, 2010) was applied and adapted to the Elba context, with the organization of a "future workshop" engaging citizens and stakeholders in future thinking. As shown in Fig. 2 below, the future-thinking approach leads participants to develop a critical view of the past, devising at the same time possible solutions in the future and finally elaborating actions for the present, having in mind strategic-long term goals.

2   The Experience in the Elba Island

The everyday—"business as usual"—scenario of mobility in the Elba island during the high season (July–August) features:

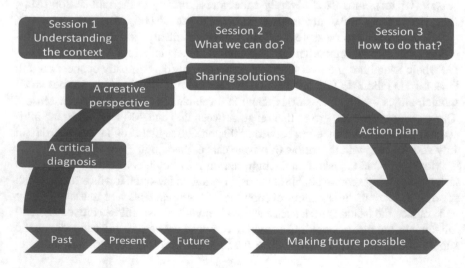

Fig. 2 Future search workshop structure

- A very high percentage of people using private vehicles for intra-island mobility (private cars are used by 80% of tourists), resulting in significant congestion levels in the summer period and the overall cutback of accessibility. The concentration of arrival/departure traffic flows in Portoferraio exacerbates congestion in the central part of the island;
- very high traffic flows crossing the centres of the different Municipalities due to the lack of bypass roads;
- a low use level of Public Transport (PT) services (only 14% of the PT offer). The mobility offer is fragmented in terms of ticketing, information, marketing, and coordination between extra-urban and urban services, failing to comply with the emerging needs of integrated solutions.

The aforementioned problems raise a wide range of issues that negatively impact the overall quality of the environment, citizens life, and tourists' experience, such as the traffic congestion in the city centres of the seven municipalities (especially in the evening), parking congestion in the areas of the most popular beaches, and road safety issues (especially for biking).

A general mobility plan for the whole Elba island would be needed to address these problems, but none has been established so far. Regulation frameworks are instead available that primarily concern the access and parking within individual municipalities, with no comprehensive, island-wide vision and policy.

However, different municipality documents recently claimed to build an "Elba SUMP" for the whole island—especially the Elba Sustainable Energy Action Plan (Patto dei Sindaci, 2014) that includes sections focused on mobility solutions.

Now, a full-fledged Elba SUMP is needed, and it should consider the main mobility challenges and measures to address the different—and somehow conflicting—travel needs of residents and tourists.

2.1 Citizens and Stakeholders' Participation

Building a vision for the whole Elba island sustainable mobility was the main ambition of the citizens and stakeholders participation process organized as one of the DESTINATIONS measures.

The Elba SUMP co-creation process was an instance of responsible research and innovation activity, aiming to align the DESTINATIONS measure—the drafting of the SUMP for the whole island—with the values, needs, and expectations of society—which were concretely represented by the citizens and stakeholders invited in the process. The interaction of science and societal perspectives is a crucial aspect of the process, as illustrated in Fig. 3 below.

Ideally, the full co-creation process should include five stages:

1. **Engage and empathize**: identify and invite local actors most affected by the topic (sustainable mobility), potentially interested to change the status quo.

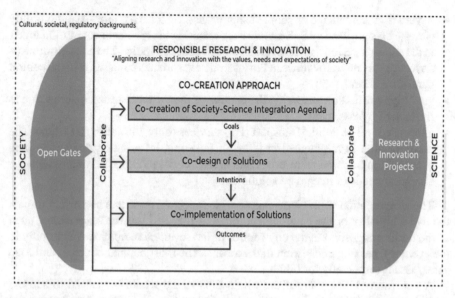

Fig. 3 Science and Society co-creation approach

They are engaged as "change agents", not in their capacity of representatives of a specific stakeholder category.

2. **Frame a shared agenda**: map the whole set of stakeholders concerned with the topic using a consistent framework to identify government, business, civil society, technical (e.g. transport operators, planners, etc.) stakeholders in place. Invite the stakeholder representatives and the change agents to a "future framing" workshop, using a Future Search methodology (Weisbord & Janoff, 2010) to structure the event. The outcome of this step is a framing agenda—the "agenda of needs for change" listing the sustainable mobility priorities.

3. **Co-design solutions**: based on the agenda of needs for change, a more detailed list of actions is sorted out and included as a key measure set in a first draft of SUMP. A "practicability" check and a fine-tuning of the most important measures are carried out with the relevant technical and administrative stakeholders (those that would be in charge of controlling and/or implementing the measure).

4. **Collaborative planning**: the measures passing the practicability check are further processed by performing a full technical, economic, normative, and financial feasibility analysis, and an ex-ante evaluation of their expected outcome and impacts against the SUMP baseline. Technical stakeholders are more heavily involved in this stage, as they would be the main actors in the next stage of implementation too.

5. **Collaborative implementation**: the final step of the co-creation process aims to anchor the collaborative work undertaken in the previous steps in the administrative process, delivering a concrete set of new policies, regulations, decisions

taken by relevant public authorities in a coordinated fashion, responding to the priorities and solutions formerly discussed with the engaged stakeholders and change agents. In the Elba case, the first collaborative implementation step would be the formal approval of a whole Elba SUMP inspired by the outcomes of the co-creation process, and its implementation and continuous evaluation and monitoring by means of a set of strategic Key Performance Indicators (KPI).

In the Elba DESTINATIONS case, the first three stages have been successfully completed, while the last two stages were beyond the scope of the demonstration activities.

The identification of the different stakeholders was an important pre-condition underlying the overall consultation and stakeholders' involvement. Considering the specific socio-economic context of the island, we have identified the stakeholder categories more involved and interested in improving the mobility situation and quality of life, environment and development opportunities for the island, as well as small groups of active citizens, the so-called "change agents":

Stakeholders: actors interested in discussing the plan and actions for sustainable mobility in the island from their perspective, including:

- Public authorities with regulatory, planning, management and control competencies over transportation in the island (Region, Province of Livorno, Municipal Administrations);
- Other public bodies with complementary skills (e.g. the National Park of the Tuscan Archipelago);
- Private and PT operators (e.g. taxi, rental vehicles agencies, Local PT operators, etc.);
- Company and/or individual technology experts and planning of mobility and services;
- Entrepreneurial associations (hoteliers, chamber of commerce, etc.);
- Civil Society and Environmental Organizations (Legambiente, etc.);
- Local and regional media.

Change agents: Individuals that share the purpose of the initiative and are interested in contributing to the formulation of the plan and the implementation of the interventions. They are asked to contribute with ideas and proposals to formulate the agenda of priorities for improving mobility and quality of life on the island. In practice, they have been organized in small groups of citizens participating in the living labs[2] organized by DESTINATIONS in the municipalities of Rio and Portoferraio.

---

[2] Living Labs (LL) are defined in a variety of ways and there is no standard definition of the concept, even though the basic idea is more or less consistent across sources. The European Network of Living Labs defines them as "a real-life test and experimentation environment where users and producers co-create innovations" (ENoLL: http://www.openlivinglabs.eu/). According to the European Commission, Living Labs have been characterized as Public–Private-People Partnerships (PPPP) for user-driven open innovation. Trying to find the lowest common denominator, a general definition is given by Bergvall-Kåreborn et al., 2009, defining them as a "user-centric innovation milieu built on everyday practice and research, with an approach that facilitates user influence in open and distributed innovation processes engaging all relevant partners in real-life contexts, aiming

Besides the above categories of stakeholders directly involved in the process, resident population, owners of second homes, and tourists are the key target groups to be considered, if the general aim of a better and more sustainable mobility is to be pursued. These target groups were represented in the composition of the "change agents" group as well, including champions of each category.

2.2   Co-creation Events

The co-creation process was implemented through a series of meetings (Future Laboratories) in various locations of the island. Meetings were upon invitation, and they always included an introduction by the organizers on the meeting purpose, followed by adequate time devoted to discussions, structured with the help of facilitators so that all participants can express their point of view and formulate proposals.

Two Future Laboratories held in Rio Marina and Portoferraio, with the participation of about 20 active citizens who are particularly interested in the theme of sustainable mobility and in the future development of the island.

Participants have used a mental map—reproduced in Fig. 4 below—to identify problems and their intervention priorities. This map helped participants to organize

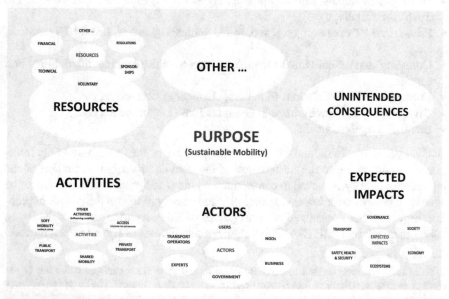

Fig. 4 Mental map of the sustainable mobility topics

---

to create sustainable values" (Bergvall-Kåreborn et al., 2009). What is striking about this definition is that it includes the goal of "creating sustainable values", focusing more on the user integration in the context of a sustainability orientation and the real-world development environment.

their collective thinking, discussing the purpose of sustainable mobility, the actors concretely affected, the activities and resources needed to achieve the purpose, and finally the expected impacts and any idea of possible unintended consequences of sustainable mobility plans.

These elements were then used to prepare a broader discussion in the Elba Future Mobility Workshop (10–11 May 2017), where the same change agents and stakeholders' representatives have been invited to participate. The activities and results of this core event are described in a workshop report (Elba Sharing, 2017), disseminated to the wider public after the event.

The main outcome of the workshop was a "shared agenda of needs and actions for the future of mobility". The workshop report also includes the full list of 30 participants, a combination of "stakeholders" and "change agents" engaged in a lively—and for most of them unusual—dialogue.

As a follow-up, the co-creation process continued after the core event to analyse the practicability of a new PT service re-organization, to meet seasonal and touristic demand. A "practicability study" was conducted involving the public authority (Osservatorio della Mobilità—Provincia di Livorno) and PT operator (CTT-Nord) in a number of meetings where the hypothesis of new bus service operation scheme was discussed in depth.

According to the new hypothesis, broadly the same volume of PT supply (measured in terms of seat.kms) would be redistributed with a grid of bus lines and services scheduled in connection with the ferries, a backbone fast and frequent service connecting Cavo in the extreme East to Marina di Campo in the West of the island, passing through the central town of Portoferraio, and a number of feeder lines to distribute passengers in the peripheral towns and attractive places (e.g. beaches, cultural, and naturalistic spots).

The practicability study—based on the analysis of origin-destination trips for leisure and touristic purpose, in addition to those for traditional purposes (school, work) already considered for planning the service during the schooling season (September–June)—has shown the convenience and economic feasibility of transforming the separate PT lines that during the winter connect peripheral municipalities to the central town (Portoferraio) into a web of integrated lines and PT rides connecting towns and attractive places across the island. More details are provided in the following Sect. 2.5. Figure 5 shows the future workshop and follow-up meetings venues.

2.3 Shared Agenda for Action

In all the future laboratories described above, the sustainable mobility topic has been introduced and contextualized to the Elba situation highlighting two macro-goals:

- Increase the number of tourists throughout the year (not only during summertime), offering transport alternatives to private car use within and for access to the island.

Fig. 5 The Elba experience: future mobility workshop and follow-up meeting

- Reduce congestion in the summer months.

Having these macro-objectives in mind, stakeholders and citizens have developed a shared "agenda for changes" deemed necessary and feasible in the short (2020), mid (2030), and long term (2050). The strategic goal is improving mobility to/from and within the island, thus also contributing to foster economic development and job opportunities, better quality of life, and lower emissions. More specifically, the agenda describes priorities for:

- Improving the access to/from and the mobility within the island favouring soft mobility (walking, biking, sailing) and shared mobility (car-sharing and pooling);

- improving the environmental performance of private mobility and enhancing the use of PT (collective taxi, shuttle services, sea cabotage, urban and non-urban buses).

This "agenda for changes" was eventually elaborated with the participants at Elba Future Mobility Workshop. Adapting the Future Search methodology, the discussions to formulate the shared agenda were structured in three steps:

- In the first step, participants identified the main factors influencing the past evolution of mobility on the island.
- In the second step, the focus shifted on the future, identifying those factors representing risks and opportunities for future mobility. Participants also identified initiatives or actions that can be accomplished—including with their direct contribution—towards the realization of the likely future.
- In the third step, participants presented conclusions and suggestions and discussed initiatives to be developed to promote the shared Action Plan.

2.4 Strategic Mission and Target Actions for the Elba SUMP

The "agenda for changes" emerged from the participatory process eventually led to defining the strategic goals and targets for the "Elba SUMP" draft elaborated as one of the measures of the DESTINATIONS demonstration in the Elba site (ISINNOVA, 2018).

2.4.1 Strategic Mission

The mission statement for the Elba SUMP (Fig. 6) is framed in one sentence: "MAKE (IT) EASY", meaning make *Elba Attractive and Sustainable all over the Year*.

The underlying key strategic objectives have been identified as follows:

1. Improving the maritime (Piombino-Portoferraio/Rio Marina/Cavo) and air (La Pila airport) access to the island;
2. Providing extensive transport information (Info-Mobility) to users and citizens;
3. Developing an integrated transport system (multimodality);
4. Enhancing e-Elba (electrical mobility);

Fig. 6 Mission statement
for the Elba SUMP

MAKE

Elba
Attractive and
Sustainable all over the
Year

5. Making the Elba island bike-friendly (bicycle mobility).

In addition, two strategic instruments were recommended as necessary initiatives to make the plan implementation concrete:

- A new strategic planning and mobility observatory in the island, with the participation of the relevant bodies (e.g. the Livorno Provincial Mobility Observatory, the 7 municipal administrations of Elba), open to a permanent consultation of stakeholders and citizens to sustain the co-creation process;
- A "Shared Mobility Agency", supporting networking among mobility service providers, info-mobility services with dynamic data to inform users in real time, and the aggregation of mobility demand allowing for sharing car rides and/or services (e.g. sharing taxi rides, blablacar, hitchhiking certification, call for flexible service that can be activated for a minimum number of users, etc.) to residents and tourists.

2.4.2 Specific Target Actions

The Elba SUMP draft includes as specific targets the actions identified at the outset of the participatory process, in the shared agenda for action. These target actions are presented below:

- **Soft mobility (walking, cycling, by sea)—actions**:
  Enhance the use of bicycles through the set-up of a network of unique off-road cycling lanes across the island.
  Encourage small cabotage to connect the island beaches.
  Sea connection between island ports with new generation boats (solar panels).
  Infrastructure and recharging network for electric bicycles and scooters at inland sites of interest (cultural sites, hospitals, beaches).
  Set-up of bike-sharing services.
- **Private and shared mobility—actions**:
  Encouraging the use of electric vehicles (free parking, dedicated stalls, charging points, vehicle access restrictions to conventionally fuelled vehicles in urban centres).
  Reconsidering parking policy in the inland: favouring intermodality, improving enforcement.
  Providing car-pooling incentives.
  Renewing transport signalling and information.
- **Local Public Transport—actions**:
  Set-up of an application encompassing all mobility services in the inland.
  Improve information on PT (timetables, GPS, info-trolleys).
  Provision of new bus lines and use of eco-sustainable transport means.
  Promotion of on-demand bus service.
  Set-up of an application encompassing all taxi services in the inland.
  Introduction of collective taxis for inter municipality routes.

Introducing additional passenger transport services during summer-time between St. Giovanni and Portoferraio.

- **Accessibility—actions**:

Improve the coordination between transport companies (timetables, local multi-modality services, integrated tickets).

Enhance transport services between Campiglia (Rail station) and Piombino (Port) in Tuscany.

Ensure better accessibility between Pisa airport and Piombino.

Encourage multimodality and interchange in Piombino Port for connections to Elba.

Incentivize the provision of private transport services, e.g. mini-cruises for coast to coast trips by sea.

Improve connections by air.

2.5 Shaping Public Transport on Elba to Meet Seasonal and Touristic Demand

Finally—as anticipated above—the main issue of the Elba SUMP that was further investigated in the co-creation process to further understand its practicability was the reorganization of the bus services (extra-urban lines) throughout the island in the summer season, to meet seasonal demand and the needs of tourists and residents alike.

The current transport network consists of a number of bus lines connecting towns and villages. Yet there is no coordination of their schedules with ferries coming into the island's three ports; Portoferraio; Rio Marina; and Cavo. This is particularly problematic during summer, the time of year when the island welcomes most of its visitors. New bus services are thus proposed for the touristic high season that would be coordinated with ferry arrival times. Whilst approximately the same volume of services would run, journey times and connectivity would improve significantly.

As shown in Fig. 7 below, a fast bus line traversing most of the island would form the backbone of this new system (main mobility corridor). Travelling from Cavo in the north-east to Marina di Campo in the west, it would pass through Elba's main town Portoferraio on the way. From the main corridor, a number of feeder lines would take passengers to peripheral towns and nearby tourist locations in 7 feed basins—the so-called *Ambiti di Mobilità Sostenibile* (AMS).

Turning the current bus connections from Elba's peripheral areas to Portoferraio into a single integrated, transport network, would thus transform mobility on the island for everyone's benefit.

To assess the practicability of the new scheme, a detailed analysis has been carried out, simulating a new organization of extra-urban lines and timetables for the summer season, summarized in Table 1.

Taking stock of the data provided by the local transport operators CTT-Nord, the number of stops, the distance (km) between stops, supply (seat.km), and demand

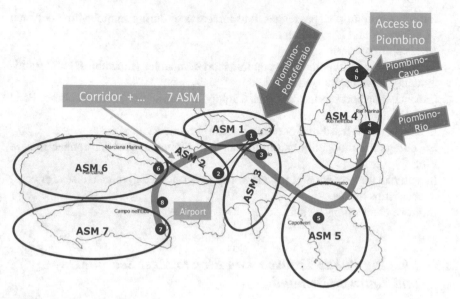

Fig. 7 Scheme of high season PT services

(passenger rides for different assumptions of tourist trip behaviours) indicators have been calculated.

Table 2 below compares the current mileage with the one emerging from the reorganization of transport services. It can be observed that the re-organization of transport services would determine a decrease in the total kilometres travelled, thanks to the rationalization of the routes, with a total reduction of 12,882 km, corresponding to a saving of €26,044 (evaluated at a direct cost of €2.021 per km).

In addition to the rationalization of the routes, the new scheme also entails an increase in the demand for PT: e.g. establishing timely connections between the bus schedules and the ferries (and flights) departing and arriving from/to the island.

The following map (Fig. 8) shows the main nodes of the network in the new scenario and for each node the specific factors that can induce an increase in the demand of PT (currently, the utilization rate of the road PT is indeed quite low).

3 Conclusions: Main Lessons Learned from the Elba Experience

Several useful lessons can be drawn from the experience of policy co-creation in the Elba island case. The co-creation approach was a success, insofar as it delivered a wealth of ideas and a coherent agenda to shift the mobility in the high season from the current unsustainable pattern towards a sustainable one. The first three steps of

Table 1 Organization of extra-urban bus services

	Current organization	Hypothesis of bus lines for the high season
Line 116	• Circuito Campo Elba • 12 bus rides Portoferraio-Marina di Campo-Pomonte-Marciana-Portoferraio	No changes envisaged **16 bus rides** Portoferraio-Marina di Campo (30 min duration). Time tables coordinated with ferries to Portoferraio **10 bus rides** coordinated with rides to Procchio and Marciana-Pomonte-Marina di Campo (90 min duration) 10 **bus rides** coordinated with rides at Marina di Campo to Pomonte-Marciana-Procchio (90 min duration)
Line 117	• 7 bus rides Portoferraio-Cavo • 3 bus rides Portoferraio-Capoliveri-Cavo	**12 bus rides** Portoferraio-Cavo (75 min). Time tables coordinated with ferries to a Portoferraio, Rio Marina, Cavo **12 bus rides** coordinated with rides at Bivio Mola to Capoliveri (10 min)
Line 118	• 5 bus rides Portoferraio-Bagnaia • 6 corse Portoferraio-Lacona	**12 bus rides** San Giovanni-Lacona (via Colle Reciso)-Bivio Lacona-Bivio Bagnaia-Bagnaia (35 min). Exchange parking at San Giovanni. Possible extension to Nisporto Coincidence at Bivio Lacona and Bivio Bagnaia with rides of the corridor Portoferraio-Cavo
3006 BIODOLA	6 bus rides Portoferraio-Biodola	**8 bus rides** Portoferraio-Villa San Martino-Biodola. Exchange parking at Villa San Martino. Rides planned until sunset
3006 VITICCIO	6 bus rides Portoferraio-Viticcio	**8 bus rides** Portoferraio-Viticcio. Rides timetables until dinner time

Table 2 Mileage of extra-urban bus services

	Current situation		New scenario	
	Daily (km)	Summer season (km)	Daily (km)	Summer season (km)
Linea 116	1,946	180,293	1,664	156,378
Linea 117	910	86,139	1,016	95,510
Linea 118	366	34,877	363	34,148
Linea 3006	262	26,248	305	28,638
km total	3,484	327,556	3,348	314,674

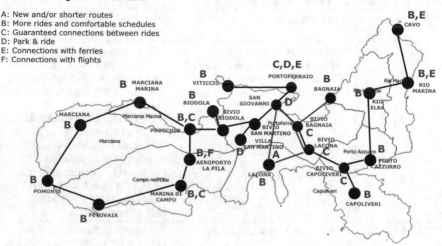

Factors inducing more PT demand:

A: New and/or shorter routes
B: More rides and comfortable schedules
C: Guaranteed connections between rides
D: Park & ride
E: Connections with ferries
F: Connections with flights

Fig. 8 Map of the high season bus network (new scenario)

the approach—engage and empathize, frame a shared agenda, co-design solutions—were successful in delivering a coherent set of goals and target actions shared by the stakeholders and the citizens engaged in the process—the "agenda of needs for change"—and led to co-create a new scenario of bus services to meet resident and tourist demand during the high season—from mid-June to mid-September and during the week-ends in the other mild weather months (March, May, October, November)—engaging managers of the regulation authority (Osservatorio dell Mobilità – Provincia di Livorno) and PT operator (CTT-Nord) in the analysis of practicability of this solution.

The further steps—collaborative planning and implementation of the solutions sketched in the draft of the Elba SUMP—were beyond the reach of the DESTINA-TIONS demonstration activities. In fact, the whole co-creation process is likely to produce the desired impact only by bringing the desired agenda and the new local PT plan to the real world of policy implementation. This would imply a further step, with the concrete commitment of the policymakers—in our case the Mayors of the 7 municipalities of the island and the regional authority in charge of mobility planning (the Province of Livorno)—to approve and launch the implementation of the SUMP for the whole island.

On one hand, we can therefore draw positive lessons about the performance of the co-creation approach and its effectiveness to frame the collaboration between different regulatory, business, and civil society actors that usually work in isolation or even distrust each other, thanks to the co-creation environment and dialogue that was helpful to combine different perspectives and achieve a mutual understanding of the needs and challenges for the future of mobility in the island. On the other hand, important lessons should be learned about what is needed as a prerequisite to really anchor the co-creation process in the policy implementation chain and deliver a concrete impact.

Perhaps the most important lesson coming out from the Elba experience—and especially the success of the living labs and the "Future of Mobility" workshop—is that different stakeholders—which know each other only superficially and often carry strong prejudices against each other—and active citizens can meet and work together very effectively to discuss and devise a common vision of what should be done—in our case delivering an agenda for sustainable mobility and better quality of life for residents and tourists. This can happen if a trustful environment and process are created by the managers and facilitators of the co-creation process. The workshop of May 10–11 2017 was an absolute novelty for most of the participants, who never met before with a common purpose in mind to assess and decide. The event helped most of the participants to understand and in some cases to change their beliefs about what is needed to improve and make the mobility in the island more sustainable.

This is encouraging. Co-creation processes as experienced in Elba should be generalized to cope with complex and multifaceted policy issues and complex problems that usually require to take decisions at different levels of government. This would entail a "governance transition", transforming a hierarchical model where the responsibility to make decisions is allocated according to the principle of subsidiarity[3]—from the lower (citizens, private sector) to the higher (local, regional, national, EU) level of government—into a "circular model" where decisions are designed, negotiated and ultimately deliberated after a co-creation process initiated by a group of "change agents" (or "visionary leaders"). The circular model engages representatives from the different public governments and agencies, the private sector, and civil society, with the support of experts to feed with technical knowledge and manage the process as appropriate. This advocated transition in the model of governance is illustrated in Fig. 9 below.

The main problem with the co-creation approach illustrated above is that its greater strength—creating a "neutral" and trustful dialogue where people with different competencies and powers are free to express their ideas and devise new solutions to common problems, without the urgency of negotiating and deliberating administrative decisions—is also potentially its main weakness. Indeed, the whole process delivers a shared agenda and roadmap for action that depends on its implementation upon decisions that should be approved and then pursued by different actors and level of governments, according to their administrative competencies and procedures.

Accordingly, the final step—anchoring the co-creation in the formal decision-making process to initiate technical planning, implementation, and evaluation of

---

[3] In a nutshell, according to this principle the competence to decide is given to the lower level of government that can solve the problem, so that solutions to small problems—without spill-over effects on others—can be decided directly by the citizens or private business themselves, while solutions that would have a significant impact at local, national, or international (EU) level shall be decided by the corresponding level of government. In practice, the solutions to wicked problems always require the competence of different levels of government, and this makes the decision process often overwhelmingly complicate and slow.

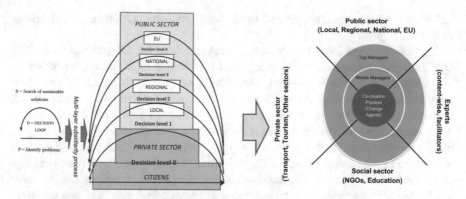

Fig. 9 Decision-making loop: from multi-layer subsidiarity to co-creation of solutions

measures—is really crucial to generate an impact, and it is often the most difficult to accomplish. For this reason, a truly successful co-creation of policies would require, to support the implementation of the full cycle of five co-creation steps we described above, a convinced endorsement and commitment of the key local and regional policymakers and/or CEOs of the agencies that play important roles in the implementation of the sustainability agenda.

As a matter of fact, a true policy commitment was present in the Elba experience only from the Mayors of the municipalities partners of DESTINATIONS—Rio and Portoferraio—and at the regional level from the Province of Livorno (Osservatorio della Mobilità) and Transport Operator (CTT-Nord). As for the latter, the participation in the co-design of the new bus services scenario was, however, limited to the technical managers, and the measures recommended in the draft SUMP plan are as yet not endorsed—at the time of writing—by any political deliberation.

This lukewarm political response is mostly due to the narrow scope of factors that drive local policymakers, who tend to focus on issues and problems affecting the single municipalities, with no real attempt to join forces to find and implement strategic solutions for the whole island. In addition, there are strong vested interests that play against any policy aiming to find alternatives to private car transport and avoid congestion in the high season: the huge and easy earnings for well-established operators in the island market—the ferries, parking lots, and other car travel chain operators—would be reduced by shifting to sustainable mobility modes. To be successful, any policy transforming the mobility in the island should therefore consider the stranded costs of the incumbent operators (primarily the ferry companies, but not only) and/or the changes needed to adapt their business practices, to find compensatory measures when appropriate.

An answer to this challenge—that could prove effective in the medium term, if not immediately—has been suggested in the shared agenda and draft SUMP plan: to gradually transform the "Elba Sharing" Agency—a DESTINATIONS measure described in another chapter of this book—in an "Elba Mobility Agency" fully participated by the key authorities—the 7 municipalities and the Province of Livorno—with competences for the regulation of new bus and shared mobility services over the whole island, also including the Port Authorities (Piombino, Portoferraio, Rio) and the airport in order to deal with the regulation of the access traffic to the island.

The whole island mobility matter would thus be administered by the Agency, taking care of mobility planning and the tendering of PT, coherently sharing services and parking. This same Agency should continuously manage co-creation processes to engage the relevant stakeholders and groups of citizens in the formulation, co-design, monitoring, and evaluation of sustainable mobility solutions.

References

Bergvall-Kåreborn, B., Eriksson, C. I., Lund, J., & Ståhlbröst, A. (2009). A milicu for innovation: Defining living labs. Retrieved September 8, 2020, from https://www.researchgate.net/public ation/228676111_A_Milieu_for_Innovation-Defining_Living_Labs.

Elba Sharing. (2017). Quale futuro per la mobilità di turisti e residenti all'isola d'Elba? Sintesi dei risultati dell'incontro del 10–11 Maggio 2017. Workshop Report (available on demand).

Kocak, N., Adell, E., Ljungberg, Ch., Ljungberg, C., Sessa, C., Giuffre, G., & Pietroni, F. (2014). Planning sustainable mobility in polycentric regions: Testing a participatory approach in six regions of Europe. *Transportation Research Procedia, 4*, 327–346.

ISINNOVA. (2018). Il Piano Elbano per la Mobilità Sostenibile – Bozza (available on demand).

Patto dei Sindaci. (2014). Piano d'Azione per l'Energia Sostenibile dell'Isola d'Elba, in collaborazione con Provincia di Livorno, Struttura di Coordinamento.

Poly-SUMP. (2014). The Poly-SUMP methodology: How to develop a sustainable urban mobility plan for a polycentric region – Guidelines, European Commission. Retrieved September 8, 2020, from https://www.eltis.org/sites/default/files/trainingmaterials/polysump-sump-guidelines-final. pdf.

Weisbord, M. & Janoff, S. (2010). *An action guide to finding common ground in organizations & communities* (3rd ed.). Berrett-Koehler.

Shared Mobility Services in Malta: User Needs and Perceptions

Suzanne Maas and Maria Attard

Abstract Malta is a small island state home to nearly half a million inhabitants. It is a major tourist destination in the Mediterranean, with visitor numbers totalling 2.6 million in 2018. There are 799 vehicles for every thousand inhabitants and the modal split shows a 74.6% reliance on the car. Ownership and use of private cars are at an all-time high in Malta. This dependence on the car has increased congestion and given rise to parking issues in many localities, taken-up limited space in the urban area, increased air and noise pollution, and created accessibility problems for pedestrians and cyclists. Shared mobility services (bicycle, car, scooter sharing) have been introduced in Malta very recently. Malta is one of the case study sites in the CIVITAS DESTINATIONS project, which focused on sustainable mobility in tourist destinations. As part of the project, a survey was conducted to understand the awareness and acceptance of these shared mobility services by Maltese residents ($n = 1,100$). Insights from this survey used to understand user needs and perceptions in light of such mobility innovations and what might encourage people to start using them as an alternative to private car use.

Keywords Shared mobility · Travel behaviour · Bicycle sharing · Sustainable mobility · Car sharing

1 Introduction

Malta is an EU member state with a land area of 316 km² and was home to a population of 460,297 in 2016 (NSO, 2018). Malta lies in the middle of the Mediterranean Sea and includes the main island Malta, Gozo and Comino. Car ownership and use are at an unprecedented high level with 799 cars per thousand (NSO, 2017) and private car use topping the modal share with 74.6% of all trips (Transport Malta, 2011).

S. Maas (✉) · M. Attard
Institute for Climate Change and Sustainable Development, University of Malta, Msida, Malta
e-mail: suzanne.maas@um.edu.mt

M. Attard
e-mail: maria.attard@um.edu.mt

Fig. 1 Valletta, the capital city of Malta. Steep streets characterize the city (left; photo by Herbert Frank, licensed under CC BY-ND 2.0), built on a peninsula between two natural harbours: the Grand Harbour and Marsamxett (right; photo by Dion Hinchcliffe, licensed under CC BY-SA 2.0)

Despite all this, the average trip distance is just 5.5 km (Transport Malta, 2016a). As other Mediterranean coastal cities, Malta's capital Valletta, as well as the broader conurbation around the harbour areas, is characterized by its historic centres and a tight spatial urban fabric that is poorly suited to the needs of car-based mobility (Cavallaro et al., 2017) (see Fig. 1). The transport system and infrastructure are under pressure from the heavy reliance on the private car, and in addition to the daily movements of a medium-sized city, such as for work, education and leisure, the transport system has to provide for seasonal tourism, especially during the summer months. As a result of high car dependence, there are pressing issues related to traffic, congestion, parking, noise and air pollution, and overall accessibility concerns for active road users. In a special Eurobarometer survey on attitudes of Europeans towards urban mobility, Maltese citizens were the most likely of all respondents in the Member States to highlight road congestion (97%), noise pollution (92%) and air pollution (95%) as important challenges (European Commission, 2013). However, due to Malta's small size and highly urbanized morphology, it may be more sensible to compare such statistics with other major cities, rather than with other European countries (Transport Malta, 2011). In Malta, transport contributes to over 20% of CO_2 emissions (Attard et al., 2015). Encouraging alternative modes of transport is not only beneficial for the well-being of local residents, in terms of reduced pollution, traffic and congestion, but can also increase the attractiveness of a destination for tourists and contribute to a better tourist experience (Le-Klähn & Hall, 2015). Shared solutions in transport, for example, (electric) bicycle and car sharing, are example of transport alternatives put forward to mitigate against the external costs of traffic and improve mobility and accessibility.

2 Sustainable Mobility and Sharing Services

The avoid-shift-improve approach is often used to guide sustainable mobility policy (Ang & Marchal, 2013; Jonuschat et al., 2015). It provides a list of priorities for sustainable mobility. Firstly, by avoiding the need to travel far by investing in a mix of land uses at the local scale and create more dense developments, and by adopting technology to substitute trips through online communications and tele-conferencing. Secondly, by shifting trips to green alternative modes, such as walking and cycling, promoting the use of public transport and encourage shared mobility. Lastly, by improving vehicle and fuel technologies and using electric, hybrid or hydrogen-fuelled vehicles which pollute less. The concept of shared mobility can be positioned as a manner that shifts mobility from individual travel to shared travel, increasing efficiency and reducing consumption. Shared mobility has the potential to provide flexible, short-term, as-needed access to mobility, as an alternative to private transport, or can be used for last-mile trips in conjunction with public transport, or as part of a multi-modal solution (Jonuschat et al., 2015; Shaheen & Chan, 2016). The widespread uptake in smartphones, combined with GPS positioning technology, increased Internet coverage, and the use of social networks has enabled people to connect, share and exchange information in new and easier ways, and has enabled the growth of shared mobility services (DeMaio, 2009; Shaheen & Chan, 2016). Shared mobility is generally understood to refer to both the sharing of a vehicle and the sharing of rides (Shaheen & Chan, 2016). In this chapter, the focus is on shared vehicles: (e-)bicycle and car sharing.

2.1 Bicycle- and Car-Sharing Systems

Bicycle-sharing systems can be classified as station based or dockless. In the former, users can rent and return bicycles from docking stations, generally from any station (although there are systems where bicycles can only be returned to the station they have been rented from). The latter refers to free-floating systems without fixed stations. Here users can find and rent a bicycle using a smartphone app within a well-defined, geofenced area (DeMaio, 2009; Fishman, 2016). Car sharing can take different forms, either in a fixed, station-based car-sharing system, where the car is checked-out and checked-in at the same location; in a flexible, free-floating system, a one-way car-sharing system in which you can book a car through an app from any location within a certain area and through peer-to-peer sharing, whereby private individuals can rent a car from another private individual through a transaction on an online platform (Jonuschat et al., 2015). Some car- and bicycle-sharing providers are now also providing additional transport options, such as shared scooters, as well as electric vehicles and bicycles (Shaheen & Chan, 2016).

Benefits of shared mobility services over private vehicles exist on different levels, for the individual, the transport operator and/or authority, and society as a whole

(Shaheen et al., 2010). For users, shared mobility can offer on-demand, flexible and increased mobility options, travel time and cost savings, and health benefits from physical activity while using bicycle sharing (Ricci, 2015; Shaheen et al., 2010). From the operator or local transport authority's perspective, shared mobility systems can provide a more efficient and low-cost alternative to investing in infrastructure for private vehicles or increased capacity for public transport (Castillo-Manzano et al., 2015; Shaheen et al., 2010). For society, benefits range from reduced congestion and improved environmental quality (reductions in air pollution and carbon emissions) resulting from mode shift and reduced private vehicle ownership (Martin & Shaheen, 2011; Shaheen & Chan, 2016), to public health improvements as a result of increased physical activity by the population (Woodcock et al., 2014; Fishman, 2016).

2.2   Shared Mobility in Malta

Malta's National Transport Strategy 2050 promotes active travel and shared mobility which are some of the guiding principles to encourage a modal shift towards sustainable transport modes (Transport Malta, 2016b). Transport Malta put forward the idea of introducing car and bicycle-sharing services in their Transport Master Plan 2025, which contains a set of measures for the shorter term. In this policy, bicycle sharing is put forward as an alternative means of travelling in the busy urban centre around the capital city Valletta, and is expected to normalize cycling in Malta for both local residents and tourists. Car sharing is suggested as a solution to improve the efficiency of vehicle use, by sharing a fleet of cars instead of relying on private vehicles (Transport Malta, 2016a). In the last few years, (e-)bicycle- and car-sharing services were introduced in Malta. Figure 2 shows the geographic distribution of the station-based (e-)bicycle- and car-sharing services (as per April 2019). Other shared mobility services, such as free-floating electric (moto)scooters, as well as ride-pooling and sharing services, have been introduced since, but were not present at the time of this research and were therefore not considered in the surveys.

The first shared mobility service to be introduced on the islands was the bicycle-sharing system implemented by *Nextbike Malta*. It was introduced in 2016, with a total of 58 stations and just over 400 bicycles. Many of the stations are located in the Northern Harbour area to the north of Valletta. This area also houses the towns of St. Julian's and Sliema which are tourist and entertainment centres, the residential and employment centres of Msida and Gżira, as well as the campus of the University of Malta in Msida. Pay-as-you-go fees for casual users are €1.50 for the first 30 min and €1 for every consecutive 30 min. Subscribed users, with a weekly (€15), monthly (€25), quarterly (€35) or yearly (€80) membership, can use the system for free for the first half-hour, after which they pay the regular fee. Stations in St. Paul's Bay also offer electric bicycles, even though the cost for these is higher (€3 for the first half-hour, €2 for every consecutive half-hour) (Nextbike Malta, 2019). Malta's public transport operator introduced the *Tallinja Bike* system in Valletta in 2018. The system consists of 40 electric bicycles at three stations (central bus station, ferry

Shared mobility services on the Maltese Islands

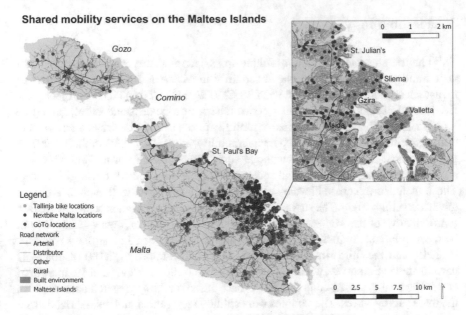

Fig. 2 Distribution of shared transport services in Malta (as in April 2019)

landing site and Fort St. Elmo), with a further two stations planned for the city's main square and at another ferry landing site. The rates are €3 for the first 30 min and €2 for every consecutive 30 min (Malta Public Transport, 2019).

In 2018, *Car2Go* put into operation a nationwide car-sharing service (called *GoTo*), following a call for proposals issued by Transport Malta (the transport regulator) and the signing of a concession agreement for the provision of services. The fleet consists of 150 electric vehicles available on both islands of Malta and Gozo and more than 300 reserved parking spaces. As a member, users benefit from 400 driving minutes at €99/month, and a rate of €0.25 per additional minute, whereas pay-as-you-go rates are €7 per month + €0.28 per minute (Car Sharing Services Malta Ltd, 2019). The car-sharing scheme *GoTo* operates across the country, while bicycle-sharing system *Nextbike Malta* is concentrated in the main urban conurbation, including some of the major tourist destinations in Malta, and e-bicycle-sharing system *Tallinja Bike* is solely available in Valletta. The shared mobility sector in Malta continues to grow, with extensions being planned by the existing shared mobility services and new scooter (moto), e-kickscooter and ridesharing services being introduced. The two available bicycle-sharing systems, *Nextbike Malta* and *Tallinja Bike,* are available with a pay-as-you-go tariff, enabling use by tourists and visitors. Following feedback from users, the car-sharing provider *GoTo* also introduced a basic plan without a membership, with a slightly higher pay-as-you-go tariff to enable casual use, as well as a pre-paid plan specifically geared at tourists.

3   Methodology

This chapter presented the results of three cross-sectional surveys to understand user needs and perceptions about shared mobility in Malta and analyse the impact of measures implemented within the H2020 CIVITAS DESTINATIONS project. This EU-funded project brings together six Southern European island cities, including Valletta, and is focused on sustainable mobility pilots in islands' urban areas to meet both residents' and visitors' transport and mobility needs (CIVITAS Initiative, 2013).

A repeated cross-sectional telephone survey was carried out in three waves to assess the people's awareness and acceptance of the new shared transport services. These included (electric) bicycle sharing and car sharing, as well as their attitudes towards cycling safety. The surveys, with a different sample in each survey wave, are representative of the Maltese population. The first wave of surveys was carried out between July and August of 2018, the second wave between January and February of 2019, and the third wave in April 2019. A total number of 1,100 respondents participated in the survey, which resulted in a 95% confidence level and 3% margin of error. The results presented in this chapter are based on the aggregated datasets from the three survey waves. The samples were split by age, gender and residential district to ensure a proportional representation of the population using a stratified random sampling strategy (Bryman, 2016). To counteract non-response, a larger sample was defined before commencing with the data collection process. Individuals over the age of 18 were invited to participate in the surveys through the use of fixed landline telephone numbers obtained from the online directory of the islands' main fixed-line provider.

The surveys contained four separate sections with 19 questions in total. Five questions related to the respondents' socio-demographic characteristics, including age, gender, locality of residence, education and employment status. Seven questions focused on awareness and acceptance of (e-)bicycle sharing and six questions on car sharing. There was only one question which focused on cycling safety. The data analysis was carried out through the use of descriptive statistics and correlation analysis using MS Excel and the open-source software R.

4   Results

The socio-demographic profile of the surveyed population is presented first and compared to the general population. After that, the awareness and willingness to use bicycle and car sharing are discussed, as well as the reasons provided for the respondents' positive or negative interest in using these services. Encouraging and discouraging factors, including the perceptions about cycling safety in relation to bicycle sharing, are then discussed.

The socio-demographic characteristics of the aggregated survey samples are presented in Table 1. The sample ($n = 1,100$) is representative of the Maltese popu-

Table 1 Survey sample characteristics

Demographic characteristics		Sample specifics($n =$ 1,100)%
Gender	Female:	52.7
	Male:	47.3
Age	18–24:	13.6
	25–34:	12.4
	35–44:	17.0
	45–54:	15.1
	55–65:	19.0
	65 + :	22.9
Employment status	In full-time employment:	39.8
	In part-time employment:	5.2
	Housewife/ Househusband:	18.7
	Retired/Pensioner:	25.4
	Student:	8.4
	Unemployed:	2.5
Highest education level	No school:	0.4
	Primary:	20.1
	Secondary:	12.6
	Post-secondary:	51.6
	Tertiary:	15.3

lation in terms of age (based on 2016 data; NSO, 2018) and gender (49.7% female, 50.3% male, in 2016; NSO, 2018), with a 95% confidence level.

The survey sought to identify the level of awareness and acceptance of bicycle and car sharing. Here, awareness is understood as being able to define bicycle and car sharing correctly (as bicycle/car rental for a period of time). Acceptance is defined as a positive response to the question 'Would you consider making use of bicycle/car sharing?'. Figure 3 shows the aggregated results of the respondents' awareness and acceptance of bicycle and car sharing, from the repeated cross-sectional surveys with Maltese residents.

Chi-square tests of the aggregated dataset were used to assess the correlation between socio-demographic characteristics (age, gender, employment and educational level) and bicycle sharing awareness and acceptance. All socio-demographic characteristics, except for gender, are significantly associated at the 95% confidence level. From the Pearson residuals of the correlation between 'age' (in age groups as presented in Table 1) and 'awareness', it becomes clear that there is a strong positive relationship between younger age groups (most pronounced in the youngest group, 18–24) and the correct answer ('bicycle rental for a period of time'), and between

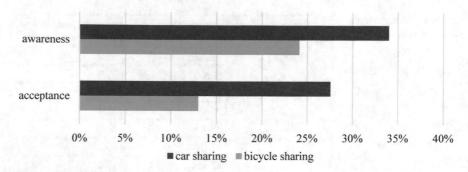

Fig. 3 Awareness and acceptance of shared mobility services by Maltese residents ($n = 1,100$)

older age groups (most strongly in the oldest group, 65+) and the answer 'don't know', and vice versa (a negative association between young respondents indicating not to know, and between old respondents and the correct answer). Figure 4 visualizes the Pearson residuals of the correlation between 'age' and 'awareness', with positive residuals (indicating a positive relationship) in blue and negative residuals (indicating a negative relationship) in red. The same pattern can be observed for their willingness to use bicycle sharing, the 'acceptance'. In terms of employment status, respondents who are 'students' and 'full-time employed' are most likely to know what bicycle sharing is and consider using it, and respondents who are 'retired/pensioner' and 'housewife/husband' are the least likely.

Awareness and acceptance of bicycle sharing are also strongly correlated with education level: those with no school and those who have attended only primary school are the least likely to know what bicycle sharing is and consider using it, and those with post-secondary and tertiary education are the most likely. A positive association between bicycle-sharing use and higher education level and (full time) employment has been found in a number of studies analysing bicycle-sharing

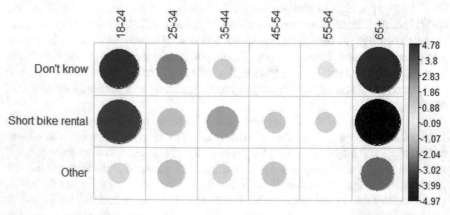

Fig. 4 Pearson residuals for the relationship between 'age' and 'awareness' of bicycle sharing

use (Fishman, 2016). The association between age and a positive attitude towards bicycle sharing are also consistent with the literature, which shows considerable uptake of shared mobility services by 18–34-year-old groups (Fishman et al., 2015; Shaheen et al., 2012). Similar results were found for the 'awareness' and 'acceptance' of car sharing in Chi-square tests assessing the relationship with the respondents' socio-demographic characteristics, as shown for the relationship between 'age' and 'awareness' in Fig. 5. 'Age', 'employment status' and 'education level' were found to be associated (at a 95% confidence level) with car-sharing 'awareness' and 'acceptance'. Gender was not found to be an influencing factor. The influence of 'age' and 'education level' follows the same patterns as observed for bicycle sharing: younger versus older age groups, and higher versus lower education have a positive versus negative influence on the awareness and acceptance of car sharing. The results for the influence of 'employment status' show a slightly different picture: respondents who are 'retired/pensioner' or 'housewife/husband' are also here least likely to have a positive attitude, but those in employment, whether 'full time' or 'part time', are the most likely to be aware of and consider using car-sharing services, whereas there is no association with respondents who are 'students' and 'unemployed'.

Tables 2a-c presents the results of the respondents' responses as a follow-up to their answer to whether they were willing to use bicycle or car sharing or not (their 'acceptance'). Respondents who replied positively were asked why they considered using bicycle/car sharing (Table 2a). Respondents who replied negatively were asked why they did not consider using the sharing service (Table 2b) and also what might encourage them to use bicycle- or car-sharing services (Table 2c).

Primarily, using bicycle sharing was seen as a form of exercise, producing less traffic and less pollution, whereas for car sharing the reasons were mainly for convenience and time savings, especially when related to parking at busy locations. Respondents seem less convinced about the promise of financial savings from the use of car sharing. The main reasons provided for the lack of use of bicycle sharing are related to knowledge on how to ride a bicycle, a preference to use private transport and the

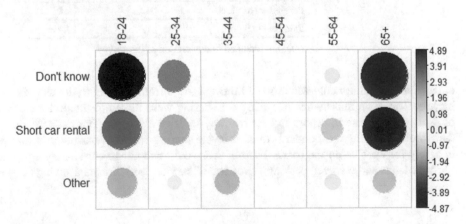

Fig. 5 Pearson residuals for the relationship between 'age' and 'awareness' of car sharing

	Bicycle sharing	Car sharing
Table 2 (**a**) Reasons for considering using sharing services (**b**) Reasons for (not) considering using sharing services and (**c**) Factors that would encourage respondents to consider using sharing services	**(a)**	
Yes, would consider using:	$n = 143\%$	$n = 302\%$
Less pollution	41.3	–
Easier parking	22.4	–
Less traffic	41.3	–
Saving money	18.2	21.9
For exercise	42.7	–
Saving time	9.8	39.1
For convenience	11.2	56.6
(b)		
No, would not consider using:	$n = 957\%$	$n = 798\%$
Price	2.0	4.6
Waste of time	9.0	10.
Prefer private transport	28.1	44.5
Family commitments	23.1	12.8
Don't know how to cycle	43.1	–
Don't drive	–	34.0%
(c)		
What would encourage you?	$n = 957\%$	$n = 798\%$
Tax incentives	2.1	5.4
Free use of service	15.2	24.8
Priority lanes, reserved parking	16.1	21.7
Increased safety	33.1	–
Financial savings	–	15.3
P&R, multi-modal connectivity	4.0	6.5
Parking restrictions	4.5	8.0
Paid parking	1.4	1.4%

Note:-not included in this section of the survey

other family commitments which would prevent users from shifting to the use of a bicycle. The main reasons for not using car sharing were linked to the preference of using own private transport and the inability to drive a vehicle. The main factor that would encourage respondents to use bicycle sharing was the possible increase in safety for cyclists on the roads. Factors that would encourage car sharing were 'priority lanes and reserved parking', 'free use of service' and 'financial savings'.

5 Discussion

While an effort was made to ensure the survey sample was representative of the islands' population, the approach used gives rise to certain limitations and potential bias. Conducting surveys using fixed-line telephony is an affordable and effective way to approach a population. However, approaching respondents through fixed-line telephone numbers only is associated with certain limitations (Bryman, 2016). There is a risk that specific groups of people using shared mobility services are under-represented. These include the younger and more technology-savvy generations who rely more on mobile phones, the increasing number of foreign (temporary) workers in the IT sector in Malta, and living in rental apartments without a fixed landline. To overcome this, sampling from both a registry of mobile and landline phone numbers (see, for example, Therrien et al., 2014) would have provided for a more diverse sample. Furthermore, while the survey sample included respondents from all over the islands, the bicycle-sharing system is geographically limited and primarily available in the urban centre of the main island Malta. Therefore, a large percentage of those participating in the survey were unlikely to have been exposed to the systems at all. Socio-demographic characteristics, such as age, employment and level of education, have a significant influence on awareness and acceptance of bicycle- and car-sharing services. In the future, campaigns surrounding this topic could narrow their focus on specific target groups that are more likely to use bicycle or car sharing, based on characteristics such as age and place of residence, while ensuring respondents represent a variety of employment and education levels.

The main factor that would encourage further take-up of bicycle sharing is increased safety for cyclists. This is in line with findings from a study in Drama (Greece), where respondents reported the lack of safe cycling infrastructure (49.1%) and limited road safety (43.6%) meant they do not consider cycling as a travel option (Nikitas, 2018). The importance of dedicated cycling infrastructure in promoting (shared) bicycle use is evident from the experiences of other cities in Southern Europe. In Seville (Spain), cycling trips increased from 10,000 to 70,000 trips per day between the period 2006 and 2011, following the creation of a 120 km bicycle network and the implementation of bicycle sharing in 2007 (Marqués et al., 2015). The introduction of bicycle sharing in 2018 and the extension of the cycling network in Las Palmas de Gran Canaria (Spain) led to 24,000 registered users after the first year of operations (Maas et al., 2020). The creation of segregated cycling paths on main roads, a reduction in speed limits in residential and rural roads, and the raising of awareness about road safety among all road users have established strategies that help to improve cycling safety (Handy et al., 2014; Heinen et al., 2010). The survey results also unearthed the relatively high percentage of people who do not know how to cycle: 37.5% of respondents in Malta in comparison with 4.7% of respondents in Drama (Nikitas, 2018). Promoting cycling training across all ages and the imple-mentation of safe routes to school for children can help improve these figures (Handy et al., 2014; Pucher et al., 2010). In fact, Nextbike Malta offers free cycling training through their Bikeability programmes held regularly in different parts of the island.

However, to date, there is limited dedicated cycling infrastructure (Transport Malta, 2016a).

The survey results show that respondents' main reasons for using car sharing are convenience and time savings. Indeed, respondents who currently do not use car sharing identified the need for priority in traffic and more reserved parking, free (subsidized service) and financial savings as factors that would change their attitude towards these new schemes. Increasing the network of shared priority lanes (reserved for shared and green vehicles) can increase the competitive advantage of shared vehicles, in comparison with the travel times and financial costs that come with using a private car. Survey results show how respondents are not convinced that a shared car may save them money. A campaign to explain the true costs of private car ownership and use when compared to a 'shared' car may be needed to raise further awareness. To encourage a modal shift, it is important to make alternatives available, but also to introduce restrictions on private car use, physically (e.g. by introducing parking restrictions or limiting access) and financially (e.g. by introducing congestion charges or parking fees).

6 Conclusions

This chapter examined the levels of awareness and acceptance of shared bicycle and car mobility services in the islands of Malta, through the aggregated results of cross-sectional surveys carried out over time, as well as the factors impacting the willingness to consider using such services, both in positive and negative terms. Younger, highly educated and full-time employed and students are main groups to target for shared mobility services in Malta. There is a strong positive relationship between younger age groups (18–44) and awareness and acceptance of bicycle and car sharing. A higher education level and being a student or full-time employed are also positively correlated.

There is a large share of respondents who are not yet aware or accepting of these new shared mobility services. Providing more information and educating people about the advantages of using shared bicycles or cars, compared to the current dominant private car use, can help in further promoting shared mobility services, specifically for target groups that are expected to be more willing to take up bicycle or car sharing, based on characteristics such as age and place of residence. To convince people to use car sharing, this can include information about monetary savings on travel, as well as time and financial savings through the provision of reserved parking spaces at destination and priority in traffic. Improved road safety and safe cycling infrastructure can have a significant impact on the use of bicycle sharing. Investment in cycling education and the creation of safe cycling routes can counter the lack of cycling abilities among the population and provides the necessary encouragement for the use of bicycle sharing as one alternative mode of transport, particularly for short trips.

References

Ang, G., & Marchal, V. (2013). *Mobilising private investment in sustainable transport: The case of land-based passenger transport infrastructure.* OECD Publishing.

Attard, M., Von Brockdorff, P., & Bezzina, F. (2015). *The external costs of passenger and commercial vehicles use in Malta.* Institute for Climate Change and Sustainable Development. University of Malta—European Commission Representation in Malta, Msida.

Bryman, A. (2016). *Social research methods* (5th ed.). Oxford University Press.

Car Sharing Services Malta Ltd. (2019). GoTo car sharing website. Retrieved April 16, 2020 from http://www.goto.com.mt.

Castillo-Manzano, J., Castro-Nuño, M., & López-Valpuesta, L. (2015). Analyzing the transition from a public bicycle system to bicycle ownership: A complex relationship. *Transportation Research Part D: Transport and Environment, 38,* 15–26.

Cavallaro, F., Galati, O. I., & Nocera, S. (2017). Policy strategies for the mitigation of GHG emissions caused by the mass-tourism mobility in coastal areas. *Transportation Research Procedia, 27,* 317–324.

CIVITAS Initiative (2013). CIVITAS DESTINATIONS project website. Retrieved April 16, 2020 from http://civitas.eu/destinations.

DeMaio, P. (2009). Bike-sharing: History, impacts, models of provision, and future. *Journal of Public Transportation, 12*(4), 41–56.

European Commission (2013). Attitudes of Europeans towards urban mobility. *Special Eurobarometer 406.*

Fishman, E., Washington, S., Haworth, N., & Watson, A. (2015). Factors influencing bike share membership: An analysis of Melbourne and Brisbane. *Transportation Research Part A: Policy and Practice, 71,* 17–30.

Fishman, E. (2016). Bikeshare: A review of recent literature. *Transport Reviews, 36*(1), 92–113.

Handy, S., Van Wee, B., & Kroesen, M. (2014). Promoting cycling for transport: Research needs and challenges. *Transport Reviews, 34*(1), 4–24.

Heinen, E., Van Wee, B., & Maat, K. (2010). Commuting by bicycle: An overview of the literature. *Transport Reviews, 30*(1), 59–96.

Jonuschat, H., Stephan, K., & Schelewsky, M. (2015). Understanding multimodal and intermodal mobility. In M. Attard, Y. Shiftan, (Eds.), *Sustainable Urban Transport*(pp. 149–176). Bingley: Emerald Group Publishing Limited.

Le-Klähn, D.-T., & Hall, C. M. (2015). Tourist use of public transport at destinations—a review. *Current Issues in Tourism, 18*(8), 785–803. https://doi.org/10.1080/13683500.2014.948812.

Maas, S., Attard, M., & Caruana, M. A. (2020). Assessing spatial and social dimensions of shared bicycle use in a Southern European island context: The case of Las Palmas de Gran Canaria. *Transportation Research Part A: Policy and Practice, 140,* 81–97.

Malta Public Transport (2019). Tallinja bike website. Retrieved April 16, 2020 from http://www.publictransport.com.mt/en/tallinja-bike.

Marqués, R., Hernández-Herrador, V., Calvo-Salazar, M., & García-Cebrián, J. (2015). How infrastructure can promote cycling in cities: Lessons from Seville. *Research in Transportation Economics, 53,* 31–44.

Martin, E. W., & Shaheen, S. A. (2011). Greenhouse gas emission impacts of Car sharing in North America. *IEEE Transactions on Intelligent Transportation Systems, 12*(4), 1074–1086.

Nextbike Malta (2019). Nextbike Malta website. Retrieved April 16, 2020 from http://www.nextbike.com.mt.

Nikitas, A. (2018). Understanding bike-sharing acceptability and expected usage patterns in the context of a small city novel to the context: A story of 'Greek Drama.' *Transportation Research Part F: Traffic Psychology and Behaviour, 56,* 306–321.

NSO (2017). *National household travel survey 2010.* National Statistics Office Malta, Valletta.

NSO (2018) *Population Statistics (Revisions): 2012–2016.* NR022/2018 National Statistics Office, Valletta.

Pucher, J., Dill, J., & Handy, S. (2010). Infrastructure, programs, and policies to increase bicycling: An international review. *Preventive Medicine, 50,* 106–125.

Ricci, M. (2015). Bike sharing: A review of evidence on impacts and processes of implementation and operation. *Research in Transportation Business & Management, 15,* 28–38.

Shaheen, S. A., Guzman, S., & Zhang, H. (2010). Bikesharing in Europe, the Americas, and Asia: Past, present, and future. *Transportation Research Record: Journal of the Transportation Research Board, 2143,* 159–167.

Shaheen, S. A., Martin, E. W., Chan, N. D., Cohen, A. P., & Pogodzinsk, M. (2012). *Public Bike-sharing in North America: Early Operator and User Understanding* (Rep. No. 11–19). San Jose, CA, Mineta Transportation Institute

Shaheen, S. A., & Chan, N. D. (2016). Mobility and the sharing economy: Potential to facilitate the first-and last-mile public transit connections. *Built Environment, 42*(4), 573–588.

Therrien, S., Brauer, M., Fuller, D., Gauvin, L., Teschke, K., & Winters, M. (2014). Identifying the leaders: Applying diffusion of innovation theory to use of a public bike share system in Vancouver, Canada. *Transportation Research Record, 2468,* 74–83.

Transport Malta (2011). *National household travel survey 2010.* Malta. Retrieved October 9, 2020 from https://www.transport.gov.mt/NHTS2010-Report-pdf_20120502091559.pdf-f1687.

Transport Malta (2016a). *Transport Master Plan 2025.* Malta. Retrieved October 9, 2020 from https://www.transport.gov.mt/strategies/strategies-policies-actions/national-transport-strategy-and-transport-master-plan-1343.

Transport Malta (2016b). *National Transport Strategy 2050.* Malta. Retrieved October 9, 2020 from https://www.transport.gov.mt/strategies/strategies-policies-actions/national-transport-strategy-and-transport-master-plan-1343.

Woodcock, J., Tainio, M., Cheshire, J., O'Brien, O., & Goodman, A. (2014). Health effects of the London bicycle sharing system: health impact modelling study. *BMJ, 348,* 11 (Clinical Research Ed.).

Is Cycling an Attractive Transport Solution in a Mediterranean City?

Dimitra Tarasi, Tryfon Daras, and Theocharis Tsoutsos

Abstract By exploiting the opinion of the society in a Mediterranean city, this paper identifies the mechanisms behind utility cycling and determines the aspects of using the bike in cities for daily commuting. This study focused on the assessment of biking as an alternative mode of transport within a typical Mediterranean city in Crete. Limited road safety is the principal reason that hinders 60% of respondents from cycling, and 41% from riding more frequently. The survey results are congruent with this statement, provided a significant difference in the ratios of actual recreational cyclists (39%) and those who like to ride for entertainment, but they hesitate (71%). In order to quantify this, the attitude of the bikers was grouped in two fact factors influencing: the clothing and the technological one. Based on the opinion of the local society, high priority policies are recommended, such as proper infrastructure, sensitization campaigns, and education programs.

Keywords Sustainability · Sustainable transport · Cycling · Urban infrastructure · Mobility

1 Introduction

Global urbanization is increasing rapidly (22% during the period 1960–2018) (World-bank, 2018); in Europe, almost three-quarters lives in cities (Eurostat, 2016); it is estimated by 2050 that 68% of the worldwide population to be located in cities (UN DESA, 2019).

Nevertheless, as cities are escalating, managing them becomes constantly more complicated, while concurrently, the quality of life deteriorates (Tsitoura et al., 2014;

D. Tarasi · T. Tsoutsos (✉)
Renewable and Sustainable Energy Systems Lab, School of Chemical and Environmental Engineering, Technical University of Crete, Chania, Greece
e-mail: theocharis.tsoutsos@enveng.tuc.gr

T. Daras
Stochastic Models of Tumor Growth Laboratory, School of Chemical and Environmental Engineering, Technical University of Crete, Chania, Greece
e-mail: tryfon@science.tuc.gr

© The Author(s) 2022
T. Tsoutsos (ed.), *Sustainable Mobility for Island Destinations*,
https://doi.org/10.1007/978-3-030-73715-3_6

Tsoutsos, 2018). The speed and sheer scale of the urban transformation presents formidable challenges to the modern world, with mobility being one of them. On the other hand, congestion in metropolis contributes not only to losses in the economic dimension but also to an increase in urban environmental pollution (Bharadwaj et al., 2017).

Urban transport has multidimensional (environmental, economic, social) implications (Purvis et al., 2018). So, as sustainable urban transport is gaining prominence, several articles highlight bike as a sustainable solution to some of the above problems that modern cities have. Cycling is eco-friendlier, healthier due to physical activity, requires limited public space, and is more economically attractive (Pucher & Buehler, 2017).

Nonetheless, how much is cycling a sustainable mode of transport within the city? Although there have been several studies concerning cycling, analyzing the best infrastructures to promote cycling, or examining the health and economic benefits (Doorley et al., 2020; Van Duppen & Spierings, 2013), the current literature is relatively limited on its sustainability as an alternative mode of transport within the Mediterranean cities. This paper seeks to address the question mentioned above, identify the mechanisms behind utility cycling, and determine the aspects of using the bike in Mediterranean cities for daily commuting. In parallel, the main aim of the current study is the analysis of the cultural facet of city cycling, given that the transition to urban sustainability is strongly linked with it. In a nutshell, this study set out to assess the current situation regarding city cycling and ascertain the potential barriers that prevent a Mediterranean city from being bike-friendly.

2 State of the Art

Concerning the health benefits, cycling can protect from chronic diseases (Buehler et al., 2011; WHO, 2004); 6–10% of the deaths for which non-communicable diseases are responsible (2.5–4.0 million annually) due to physical inactivity (Biddle, 2016; Kohl et al., 2012). Moreover, it plays a role in decreasing cancer risk, with a 40–50% reduction for colon cancer, a 20–30% reduction for breast cancer, and a 20% reduction for ovarian and lung cancer (Rebecchi et al., 2016). According to statistics, only a small share (17%) of the Greek population spends at least 2.5 h/week of leisure time doing physical activities, such as cycling (Eurostat, 2017); this is the minimum level of physical activity recommended by the World Health Organization for adults (WHO, 2010).

On the other hand, the environmental advantages of cycling cannot be ignored. In 2014, the transport sector was responsible for 21% of global CO_2 emissions (Worldbank, 2014). On the contrary, the bike can contribute significantly to the mitigation of climate change. More specifically, a study concluded that a dramatic increase (about 20%) in urban cycling worldwide could decrease CO_2 emissions from transportation by 11% (Mason et al., 2015). Additionally, cycling reduces air pollution, as well as water and soil contamination, alongside having a smaller manufacturing footprint.

Furthermore, replacing car by bike can achieve economic benefits both for the citizens and for the society. Cycling contributes to more livable cities via less congestion, better air quality, less noise, and more space for people. These features, in turn, create attractive metropolis, with increased tourism revenue from eco-tourists, which is crucial for the Mediterranean cities. In parallel, it saves money as it reduces expenditure on health, on vehicles, on-road surfaces, on expensive, and polluting fossil fuels (Neun & Haubold, 2016). Besides that, cycling-related industries create a series of new jobs, and investments in cycling support local shops (Dekoster and Schollaert 1999).

Last but not least, the social benefits of cycling are regarded to be equally essential. The bike provides increased mobility to low-rate social groups, such as low-income earners and youngsters (Ji et al., 2017). An additional convenience that is worth noting is the fact that it enables people to interact socially, and this enhances a sense of community. Similarly, shared cycling and pedestrian infrastructure, such as improved paths, also create benefits for pedestrians and people bringing justice between population groups. A study shows that specific infrastructure and behavioral change encourage to cycle from home to work regularly (Xu & Chow, 2020).

Notwithstanding, several cities in Europe demonstrate the feasibility of the reduction in the use of private cars. Nonetheless, despite the multi-faceted merits of cycling, many cities continue to privilege vehicles by prioritizing car-centric policies and planning (Mayers & Glover, 2019). Most Greek cities fall into the abovementioned rule, due to the urban development which lacks infrastructure; regardless the fact that there has been a boost in cyclists the last decade, as a result of the recent financial crisis, Greece is not generally recognized as a country where bike culture prevails, despite the favorable weather conditions.

It has been conclusively shown that roads which guarantee a high level of safety encourage residents to use bikes.

2.1 Study Area

According to the latest census, the permanent population of the Municipality of Chania amounts to 108.642 inhabitants (our area of interest was the municipal units of Chania downtown, Akrotiri and Souda with 61,275 inhabitants (ELSTAT, 2011)). The climate of the region is the subtropical Mediterranean, with sunny, dry summers, and very mild, rainy winters. Snow and frost are rare near the coast, and during the period between April and October, clear-sky weather is almost a common feature. Concerning the topography, the city's center can be described as mild, with minimum elevations. In contrast, high-altitude differences of over 200 m can be observed to the Akrotiri Peninsula, which extends northeast of the city. It is noteworthy to point out that the Technical University of Crete, a major part of the city's life, is located in the peninsula mentioned above. Last but not least, regardless of the lack of cycling network, in 2015, a municipal bike-sharing system was officially launched with four stations within the city's center boundaries.

2.2   Research Methodology

Cycling in the city of Chania has been scarcely investigated from the transportation point of view; most studies focus on the potential construction of cycle lanes within the city or the near suburbs. Within this framework, parameters such as the volume of cyclists–bike owners, the frequency and reasons for cycling trips, the way that citizens commute to their work or the city's center and their notion regarding the bike use as an alternative transport mode have not to be taken into consideration. That is to say, there is no prior research concerning the topics mentioned above, neither information that depicts the present situation of the city.

There were contacts with: (a) European Cyclists' Federation; (b) Confederation of the European Bicycle Industry; (c) Greece's major bike manufacturing companies which could provide the annual sales of the industry, alongside with the market´s tendency; (d) local bike dealers and shops with the view to form an opinion respecting the local market and the proportion of the population that cycle; and (e) Chania's amateur cycling group Podilatreis (in Greek: bike lovers), which was considered of utmost importance, taking into account the fact that they are active the last decade and they could conceivably provide valuable information.

Under the lack of essential cycling raw data, and in order to gain insight into the current state regarding cycling in the city of Chania, a survey has been conducted. The four-stage work was implemented with the support of a structured questionnaire (Bechhofer & Paterson, 2000; McNeill, 1990), and tested through a pilot study:

1. the development of a structured questionnaire;
2. its distribution;
3. data gathering/recording; and
4. data analysis.

The questionnaire included both close-ended and open-ended questions (Bradburn et al., 2004) and grouped into the following five sections:

1. personal data;
2. general information;
3. transportation within the city;
4. cycling in the city and infrastructures; and
5. driving conduct.

It was regarded as vital to include the behavioral aspect in the survey, as it is partially unexplored and obscure yet of great significance. Commonly, cycling research is mainly oriented toward quantitative data collection, and calls little, if any, attention to the qualitative features of the subject. Moreover, participants were allowed to state their viewpoint freely on the issue, exchange their ideas and make suggestions.

The research carried out during the period December 2018–January 2019 was addressed to adult citizens of Chania and it was distributed both online, as it ensures the reaching out to a broader population, and in the paper. The digital format of the questionnaire was created with the tool "Google forms." Notably, social media platforms enhanced further promotion of the survey. In all cases, participants' consent was obtained, and since the questionnaire was anonymous, there was no need to abide by the General Data Protection Regulation (GDPR). The collected data were analyzed using SPSS v.20 (Apostolakis et al., 2009; Green et al., 2000). Besides, they were tested for face-value validity and reliability (Cronbach, 1951).

3 Results

3.1 General Information

The random sample consisted of 272 citizens, weighted w.r.t. the percentage of current/projected (by the future sales) bike owners (53.7% men and 46.3% women). This analogy between men and women is estimated as the rationale for local bikers based on the (empirical) experience of the researchers. Of the study population, 74.3% owns a bike, and among those who do not, only one-quarter responded that it is quite possible to purchase one soon. There is a dependence relation between the sex of a survey participant and the fact that he/she is a bike owner (men are more often bike owners, 87.7–58.7% for women, $x^2 = 29.640$, sig $= 0.000$) (Table 1).

Regarding the participants' age, the majority, 36.4%, is middle-aged (35–44 years old). There is also a dependence relation between the age of a participant and the fact that he/she is a bike owner (adults of age 45–54 are more frequent owners, 90% $x^2 = 16.712$, sig $= 0.000$) (Table 2).

Concerning the utility reasons, there was no clear trend observed, given the fact that there was an absolute equilibrium between transportation and physical activity (47.5% and 47.9% respectively). In other words, for one out of two participants, the cited purpose for cycling is either traveling or exercising. On the other hand, a figure of 38.9% declared themselves as recreational cyclists.

There is a statistically significant difference in the way men and women use the bike: men usually use the bike for transportation (55.2–38.5% for women), physical

Table 1 Bike ownership w.r.t. sex

			Bike owner		Total
			Yes	No	
Sex	Man	Count	128	18	146
		% within sex	87.7	12.3	100.0
	Woman	Count	74	52	126
		% within sex	58.7	41.3	100.0
Total		Count	202	70	272
		% within sex	74.3	25.7	100.0

Table 2 Bike ownership w.r.t. age

			Bike owner		Total
			Yes	No	
Age	18–24	Count	23	20	43
		% within age	53.5	46.5	100.0
	25–34	Count	43	12	55
		% within age	78.2	21.8	100.0
	35–44	Count	73	26	99
		% within age	73.7	26.3	100.0
	45–54	Count	45	5	50
		% within age	90.0	10.0	100.0
	Over 55	Count	18	7	25
		% within age	72.0	28.0	100.0
Total		Count	202	70	272
		% within age	74.3	25.7	100.0

activity (60.1–33.6%), recreation (46.9–29.5%) and women tend not to use the bike at all (45.9–18.9 for men).

When questioned about the frequency of riding, 29.5% claimed to cycle 4–7 times/week, while 18.8% mentioned using the bike less, that is to say, 1–3 times/week. It is worth noting that almost three out of ten participants do not cycle at all, even though they are bike owners. The rest of the sample, 24.1%, alleged occasional use of 1–3 times/month or less. Men (38.5%) and adults of age range 45–54 (42.9%) are more frequent bike users (Fig. 1).

Notwithstanding, 70.2% of the individuals declared not to cycle as frequently as they would like (81.8% of women to 60.4% of men, difference statistically significant, $x^2 = 14.393$, sig $= 0.000$).

Concerning the causes that hinder the more frequent use, the lack of infrastructure prevails with a striking 70%. Nevertheless, safety reasons do not fall far behind, as they prohibit 41.4% of cyclists from riding more often. Third on the list was encountered the weather conditions since 32.7% referred to them as the main reason for infrequent use (Table 3).

The results can also be verified by employing (through exploratory factor analysis/ e.f.a.) of two latent variables/factors. One factor loads the variables "Tiring," "Time," and "Other Means" (named Factor1) and a second one loads the variables "Not Safe" and "Infrastructure" (named Factor2) (here KMO $= 0.531$, for Bartlett's test p $<$ 0.05 and factors explain 53.56% of the common variable variance) (Fig. 2). Thus, we have to keep in mind the non-existence of proper infrastructure relates (strongly) with safety concerns.

Interestingly enough, the data yielded by this study provide convincing evidence that the lack of bike lanes and safety reasons are again responsible for not using a bike. More specifically, one out of three subjects do not cycle, due to the absence of

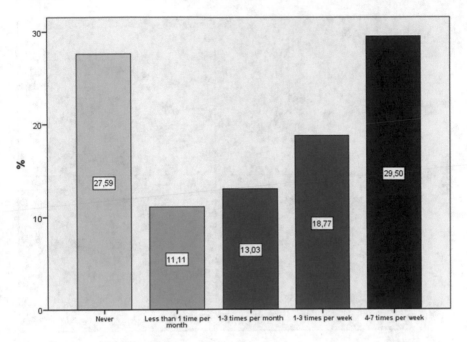

Fig. 1 Frequency of the bike use

		Responses	% of cases
Table 3 Reasons for not frequent use		N	
Reasons for no frequent use	No infrastructure	154	70.0
	Not safe	91	41.4
	Not a bike owner	37	16.8
	Weather	72	32.7
	Other means	29	13.2
	Other	14	6.4
	Tiring	13	5.9

a bike network within the city in one hand (63.2%), and safety reasons in the other (56.9%).

Regarding the major benefits of cycling, the available data indicate that almost half of the sample (47.4%) evaluated the easy transportation as an outstanding advantage. In contrast, for 30.4% of the participants, the fact that bike is an environmentally friendly means of transport with zero emissions of CO_2 is of utmost importance. Finally, there is the smallest proportion of individuals (11.9%), who underlined the fact that it is a low-cost way of traveling.

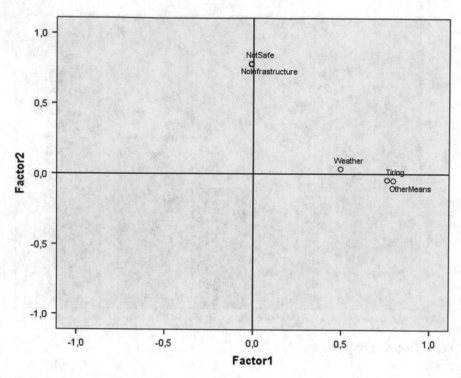

Fig. 2 Graphical representation of variables in Factors' space

There is a difference, statistically significant ($x^2 = 13.137$, sig. $= 0.004$), between men and women on which is the most crucial advantage of bike riding. Women considered as the main reason the fact that there is no environmental pollution (39.2–22.8% for men), in contrast with men who think its main advantage is that it is an easy transportation mean (57.2–36% for women) (Table 4).

Table 4 Advantage of bike use w.r.t. sex

			Easy transportation	Cost-efficient	No environmental pollution	Other	Total
Sex	Man	Count	83	16	33	13	145
		% within sex	57.2	11.0	22.8	9.0	100.0
	Woman	Count	45	16	49	15	125
		% within sex	36.0	12.8	39.2	12.0	100.0
Total		Count	128	32	82	28	270
		% within sex	47.4	11.9	30.4	10.4	100.0

3.2   *Transportation Within the City*

When a hypothetical question was posed to the subjects, in respect of the potential use of the bike, the majority of those surveyed (71.9%) indicated that they would cycle for recreation. In contrast, an almost equal proportion (70%) suggested that it would be useful for commuting to the work or the university. Besides, one out of two persons asserted that he/she would ride a bike to go shopping. In other words, it would be used as a mode of transport.

Men and women use the bike, inside the city, in different ways. Men mainly use the bike for commuting to work (64.1–47.2% for women), and women, on the other hand, use it mostly for shopping (64.8–49% for men). Additionally, concerning age, the majority of those in the age group 18–24 use it for commuting to university (40.5%). These differences are statistically significant (level of significance a = 0.05).

In another scenario, slightly over half of the study population (54.8%) would definitely choose the bike over another means of transport for traveling within the city. The same standpoint, but with an insignificant reservation, shared 22.6% of the respondents.

Men, to a greater extent than women, would use the bike over any other means of transportation (65.5–42.4% for women, difference statistically significant, $x^2 = 15.132$, sig = 0.004). The prevailing reasons for opting for a different means of transport instead of bike coincided with the previously mentioned ones. More precisely, the findings demonstrate that, for 71.6% of the total, the key factor was the lack of infrastructure, followed by 50.6% who highlighted the safety parameter. Lastly, the adverse weather conditions were the underlying argument in favor of choosing other modes of transportation for 37% of the study population (Table 5).

Moreover, these results can be verified, in a similar manner as in paragraph 4.1, with the use also of the same two factors.

Women feel less safe, when cycling, in contrast with men (59.3–41% for men, $x^2 = 5.489$, sig = 0.019).

Table 5 No preference frequencies

	Responses N	% of cases
Not a bike owner	30	18.5
Not safe	82	50.6
Tiring	12	7.4
Weather conditions	60	37.0
Infrastructure	116	71.6
Time	20	12.3
Other means	16	9.9
Other	5	3.1

Also, members of the sample of age group 18–24 (51.7%) consider less, when cycling, the absence of infrastructure (in comparison with the other age groups, they think in a percentage of more than 80% the infrastructure).

Drawing from the survey results, it is evident that infrastructure and safety were the two features of paramount importance for the participants and the key drivers of their viewpoint regarding the adoption of cycling as an alternative way of transportation. Even though indirectly it had already been underpinned the fact that the sample does not feel safe to cycle, a direct question in which they had to evaluate the level of security while cycling, accentuated it even more. To be more specific, a significant proportion, 43.7% of the people surveyed, admitted to feeling totally unsafe, while another 25.9% and 23.7% reported a high and moderate feeling of insecurity, respectively. Furthermore, women feel less safe when cycling (57.6–31.7% for men), and in contrast, bike owners feel safer (63.2–37.1% for non-bike owners).

The safety issue also emerged when the sample was requested to assess the feeling of being exposed to the possibility of theft or harassment as cyclists, given that for the majority (72.8%), this risk is high.

Nonetheless, which are the predominant challenges that citizens are experiencing when they are traveling within the city? According to 74.3%, the number one adversity they are facing is road safety. Consequently, this issue arose again, and it became crystal clear that it is a subject of great importance for the inhabitants. A high rate of the respondents (63.6%) described traffic congestion as a significant problem, while air contamination and noise pollution were mentioned by the 32.2% and 26.4% of the sample, respectively. Last but not least, a figure of 26.4% declared that the principal impediment in their everyday commuting is the delay due to congested roads (Fig. 3).

Also, using e.f.a., the main challenges/problems can be expressed using two factors. The first factor includes the variables "Traffic," "Delays" and we could probably name it "Traffic Problems"; on the second factor, the variables "Pollution" and "Sound Pollution" are being loaded, and we can call it "Pollution Problems" (here KMO = 0.601, Bartlett's test $p < 0.05$ and factors explain 67.85% of the common variable variance). The factor loadings (using Varimax rotation) of the variables can be seen in Eqs. 1–2:

$$\text{Traffic Problems} = 0.694 * \text{Traffic} + 0.515 * \text{Delays}$$
$$- 0.194 * \text{Pollution} + 0.58 * \text{Sound Pollution} \qquad (1)$$

$$\text{Pollution Problems} = -0.203 * \text{Traffic} + 0.077 * \text{Delays}$$
$$- 0.688 * \text{Pollution} + 0.520 * \text{Sound Pollution} \qquad (2)$$

So, except for safety, policy measures have to be taken by the state/municipality for traffic and pollution problems.

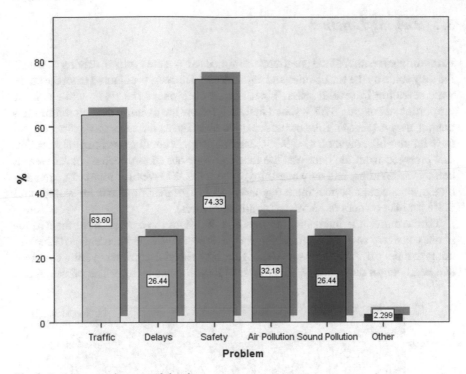

Fig. 3 Problems moving around the city

3.3 Cycling in the City and Infrastructures

Another interesting point is the citizens' opinion and reaction concerning the construction of bike lanes. What would happen if there was a complete bike network within the city? Would they opt for cycling? The results of the survey concerning this matter appear to be undeniable, provided that an overwhelming 90% claimed that, in case of infrastructure, they would use a bike. Hardly a 2.6% was opposed to this idea, while the remaining 7.4% stated: "I don't know." Bike owners are more willing to use these type of networks (96–72.1% for non-bike owners, $x^2 = 33.182$, sig $= 0.000$).

Furthermore, a similar attitude was observed, respecting the use of a bike in the case of infrastructure. When the subjects were questioned if they would cycle for commuting to their workplace or the university, an astonishing 79.2% reported "Yes," in contrast with a 12.6% who opposed to it, while the remaining 8.2% appeared indecisive.

Similarly, impressive is the proportion of those who are in favor of the construction of cycle lanes. 97% of the participants would support such an initiative, while 88.4% of them consider the existence of bike parking facilities essential.

3.4 Driving Conduct

An additional section of the questionnaire aimed to evaluate cyclists' driving conduct. Do they abide by the traffic rules and regulations? The results obtained from the study were revealing in several ways. First, almost 60% obeys the rules, while a great proportion of roughly 37.3% stated that they follow the existing driving regulations but not always (Fig. 4). Bike owners seem to obey the traffic rules more often (60.8–54% for non-bike owners, $x^2 = 9.396$, sig $= 0.009$). Secondly, a striking observation that emerged from the data was the fact that four out of ten cyclists never wear a helmet while riding, and only one-third of them (32.9%) wears it regularly. Men and bike owners wear a helmet more regularly (43.1% for men to 18.6% for women and 6.5% for bike owners to 13.5% for non-bike owners).

Furthermore, it is interesting to note that 23.3% of the cyclists have their head-phones on every so often when riding, and an almost equivalent percentage (22%) do not take them off. Also, only one out of four bike riders use proper cycling clothing, and nearly eight out of ten (84.3%) wear at least light reflectors. The percentage is

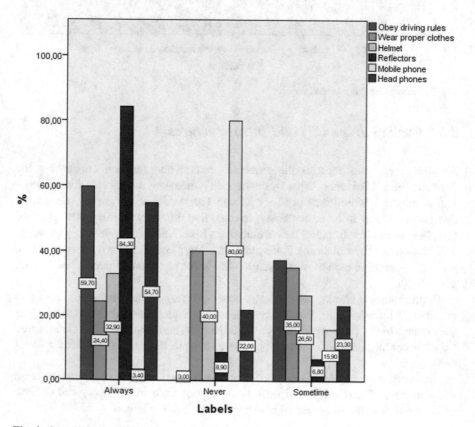

Fig. 4 Overall driving behavior

even higher for bike owners, 87.9% to 64.9% for non-bike owners ($x^2 = 14.629$, sig $= 0.001$).

As a final point, the use of a mobile phone while riding is undoubtedly a significant issue, given that this practice underlies a high risk. The findings demonstrate that the majority (80.7%) of the study population does not use it when riding, and only the minority (15.9%) uses it occasionally.

By using exploratory factor analysis the driving behavior data processed (i.e., variables "Proper Clothing Use," "Helmet Use," "Night Light Reflectors Use," "Headphones Use," "Mobile Use") in order to find out latent factors justifying the data (not measured, driving behavior, variables).

Choosing principal component analysis and orthogonal Varimax rotation, and since all KMO values for individual variables were above 0.05, Kaiser–Meyer–Olkin measure was also > 0.50 (adequacy of the sample), Bartlett's test of sphericity p < 0.05 (no multicollinearity problems), and an eigenvalue cut-off of 1.0 (also confirmed by the scree plot), we ended up with two factors explaining 58.17% of the common variance.

The variables "Proper Clothing Use," "Helmet Use" are being loaded to the first factor, through the model of Eq. 3:

$$\text{Clothing Behavior} = 0.905 * \text{Proper CLothing} + 0.905 * \text{Helmet Use} +$$
$$+ 0.162 * \text{Night Light Reflectors Use} + 0.134 * \text{Mobile Use}$$
$$- 0.113 * \text{Headphones Use} \tag{3}$$

and we can probably call this factor Clothing Behavior. Furthermore, the variables "Night Light Reflectors Use," "Headphones Use," and "Mobile Use" are being loaded to the second factor, through Eq. 4,

$$\text{Technology Behavior} = 0.012 * \text{Proper Clothing Use} + 0.034 * \text{Helmet Use} +$$
$$+ 0.681 * \text{Night Light Reflectors Use}$$
$$+ 0.611 * \text{Mobile Use} - 0.613 * \text{Headphones Use} \tag{4}$$

and we can probably call this factor Technology Behavior. The higher the individual scores w.r.t. the variables/factors, the better the load to the factor (for the reliability of the two new factors: Cronbach's alpha > 0.750).

So the (explanation of psychology behind) driving behavior could be split into two parts, the clothing equipment on the one hand and the technical equipment on the other and a (possible) future policy could be based on these two different parts. Also, using t-tests and ANOVA, there is a (statistically) significant difference in Clothing Behavior w.r.t. sex, age, and bike ownership. Men, individuals of age 34–45, and bike owners score higher w.r.t. this factor. Also, there is a (statistically) significant difference in Technology Behavior w.r.t. bike ownership.

Finally, it is worth noting that 47 respondents declared their personal perspective regarding the subject, and made beneficial recommendations and comments.

4 Discussion and Conclusions

Under the lack of bike lanes, the city of Chania can be identified as car-centric, as a consequence of traffic congestion, poor air quality, and noise pollution (as we have also seen in the factor analysis). The survey results are congruent with this statement, provided that a significant difference in the ratios of actual recreational cyclists was observed (39%) and those who would like to ride for entertainment, but hesitate (71%). Nevertheless, according to some participants' comments, car-centrism is not only preventing citizens from cycling, but it is also a cultural problem that will take much time to change.

A recurrent theme in the study which emerged, in most cases indirectly, yet of utmost importance, was a sense among participants of none or limited road safety. This is the principal reason that hinders 60% of respondents from cycling, and 41% from riding more frequently. Authors of studies in this field point to the fact that roads which do not guarantee a high level of safety discourage residents from using bikes (Van Holle et al., 2014). Enlightening is the comment of a participant: *"After the incidents of serious and fatal injuries to cyclists over the past few years, I confess that fear prevents me from getting back on the bike even though I had bought a brand new bike."*

Nonetheless, the insecurity mentioned above stems from the lack of appropriate infrastructure in one hand and poor driving conduct on the other (both verified by the exploratory factor analysis). According to the findings, there is a significant and undeniable relationship between the absence of bike lanes, and none or limited use of the bike. As one individual stated: *"No cycle lanes, no safety."* A similar philosophy was observed for 63% of the respondents, who reported a compelling reason for not cycling the lack of infrastructure. Moreover, an even higher proportion (70%) not only would cycle more if it was less risky, but also would opt for a bike instead of other means of transport.

The significance of infrastructures and the positive impact that would have on the mobility behavior of the citizens has been highlighted already by several researchers (Majumdar & Mitra, 2018). In this study, the results of the survey demonstrated that, if the better infrastructure of bike lanes was available, the outstanding 90% of the citizens would cycle. Among them, 79% would commute to work or the university by bike. Secondly, the participants' statements left no room for doubt. In other words, infrastructures would set a clear trend toward the change in travel habits and the adoption of cycling.

An additional crucial issue that emerged through this study was cyclists' poor driving skills. There were a plethora of comments which referred to this topic and accentuated the necessity of immediate improvement.

A further interesting finding of this research that it should be taken into account is the fact that the inhabitants appeared to be almost unanimously (97%) in favor of the construction of cycle lanes. Moreover, a common belief among them was the strong link between appropriate facilities and the promotion of cycling. As one individual put it: *"Only infrastructure can push citizens to use the bike."* This finding corroborates the ideas of Cervero et al. (2013), who suggested that building safe, secure, and well-designed bike infrastructure increases the number of cyclists and encourages cycling.

One rather disappointing finding was the fact that former cyclists have abandoned riding their bikes as a consequence of always feeling conscious of their safety while cycling. Without any doubt, the lack of infrastructure, in tandem with the poor driving conduct, contributes significantly to this sensation.

Overall, these results suggest that it is imperative to emphasize the construction of the appropriate infrastructure in one hand, and on the cultivation of cycling culture in the other. Alongside, it is vital to ameliorating the citizens' driving behavior.

Despite the advances toward a sustainable city, obviously, the late financial crisis that resulted in critical modifications in the everyday life of citizens, including travel behavior, Chania remains a car-centric city. The lack of prioritization for cycling is a cultural barrier, which has proven difficult to change. These changes require long periods, during which the population develops new mobility behavior. Thus, it will be a long time until the bike outweighs the vehicles regarding transportation within the city.

In order to achieve the desired mindset and meet the constant need for more sustainable mobility, the implementation of specific policies aiming to the cultivation of cycling culture is indispensable. More precisely, the use of "soft interventions," such as awareness campaigns and events, has proven to be remarkably effective. Particular attention should be given to the young generation, given that it is of paramount importance to inculcate in them the philosophy of sustainable and environmentally friendly transportation. For that reason, educational programs should be incorporated in schools' curriculum, in hand with experiential ones, because experiential learning has astonishing results. Nevertheless, cycle training for adults, who are either inexperienced or wish to regain their riding confidence, could be an additional measure. Concerning the promotion of city cycling, a series of initiatives, in particular, subvention for bike purchasing, incentives for cycling to work, and expansion of the bike-sharing project, could have as well a positive effect (Table 6).

Turning now to the future work that could be performed in this field, more extensive research could establish more robust evidence. Aside from that, taking into serious consideration the fact that studies on the behavioral change regarding cycling in the city are lacking, further research on this topic is required. Another critical question for future studies to resolve is whether people perceive the long-term benefits of limiting the use of the private car. In a broader level, research is also needed in order to determine the reasons behind the poor driving conduct.

Table 6 Recommended policies to encourage cycling in the city of Chania

Policy measures	Target audience	Actor	Time horizon
Change of the mindset			
Behavioral change educational programs	Children Students	Municipality Schools Academia	Short term
Experiential learning programs	Children Students	Municipality Bicycle clubs Parent associations	Short term
Safety campaigns	Women	Municipality Schools Academia	Short term
Environmental campaigns	Men	Municipality Schools Academia	Short term
Infrastructure			
Improvement of the quality and extension of infrastructure	General Public	Municipality State	Medium–Long term
Traffic problems' monitoring and improvement	Municipality General public	Municipality/ State	Short–Long term
Regulation monitoring	General public	Police Municipality	Short term
Incentives			
Incentives for cycling to work	General public	State Employers	Short term
Subvention for bicycle purchasing	General public	State Bicycle dealers	Short term

References

Apostolakis, I., Daras, T., & Stamouli, M. A. (2009). *Exercises in applied statistics in health sector* (Vol. 2). Papazisis Publishers (in Greek).

Bechhofer, F., & Paterson, L. (2000). *Media reviews. Principles of research design in the social sciences.* Routledge Taylor & Francis Group.

Bharadwaj, S., Ballare, S., Rohit, & Chandel, M. (2017). Impact of congestion on greenhouse gas emissions for road transport in Mumbai metropolitan region. *Transportation Research Procedia, 25*, 3538–3551.

Biddle, S. (2016). Physical activity and mental health: Evidence is growing. *World Psychiatry, 15*(2), 176–177.

Bradburn, N., Sudman, S., & Wansink, B. (2004). *Asking questions. The definitive guide to question-naire design for market research, political polls, and social and health questionnaires* (Revised). Jossey-Bass-John Wiley & Sons Inc.

Buehler, R., Pucher, J., Merom, D., & Bauman, A. (2011). Active travel in Germany and the U.S. *American Journal of Preventive Medicine, 41*(3), 241–250.

Cervero, R., Caldwell, B., & Cuellar, J. (2013). Bike-and-ride: build it and they will come. *Journal of Public Transportation, 16*(4), 83–105.

Cronbach, L. (1951). Coefficient alpha and the internal structure of tests. *Psychometrika, 16*(3), 297–334.

Dekoster, R., & Schollaert U. (1999). *Cycling: the way ahead for towns and cities*. European Commission, Luxembourg.

Doorley, R., Pakrashi, V., Szeto, W. Y., & Ghosh, B. (2020). Designing cycle networks to maximize health, environmental, and travel time impacts: An optimization-based approach. *International Journal of Sustainable Transportation, 14*(5), 361–374.

ELSTAT. (2011). *Census 2011*. Retrieved August 20, 2019, from http://www.statistics.gr/2011-census-pop-hous.

Eurostat. (2016). *Share of urban and rural populations, 1950–2050 ([1]) (% of total population) Cities16.png—Statistics explained*. Retrieved August 20, 2019, from https://ec.europa.eu/eurostat/statistics-explained/index.php?title=File:Share_of_urban_and_rural_populations,_1950%E2%80%932050_(%C2%B9)_(%25_of_total_population)_Cities16.png.

Eurostat. (2017). *How much do Europeans exercise?*. Retrieved August 20, 2019, from https://ec.europa.eu/eurostat/web/products-eurostat-news/-/DDN-20170302-1.

Green, S., Salkind, N., & Akey, T. (2000). *Using SPSS for windows. Analyzing and understanding data* (2nd ed.). Practice Hall.

Ji, Y., Fan, Y., Ermagun, A., Cao, X., Wang, W., & Das, K. (2017). Public bicycle as a feeder mode to rail transit in China: The role of gender, age, income, trip purpose, and bicycle theft experience. *International Journal of Sustainable Transportation, 11*, 308–317.

Kohl, H., Craig, C., Lambert, E., Inoue, S., Alkandari, J., Leetongin, G., & Kahlmeier, S. (2012). The pandemic of physical inactivity: Global action for public health. *The Lancet, 380*(9838), 294–305.

Majumdar, B., & Mitra, S. (2018). Analysis of bicycle route-related improvement strategies for two Indian cities using a stated preference survey. *Transport Policy, 63*, 176–188.

Mason, J., Fulton, L., & McDonald, Z. (2015). *A global high shift cycling scenario: The potential for dramatically increasing bicycle and E-bike use in cities around the world, with estimated energy, CO2, and cost impacts*. Institute for Transportation & Development Policy and University of California, Davis, USA.

Mayers, R., & Glover, T. (2019). Whose lane is it anyway? The experience of cycling in a Mid-sized City. *Leisure Sciences*, 1–18.

McNeill, P. (1990). *Research methods* (2nd ed.). Routledge.

Neun, M., & Haubold, H. (2016). *The EU cycling economy—arguments for an integrated EU cycling policy*. European Cyclists' Federation, Brussels.

Pucher, J., & Buehler, R. (2017). Cycling towards a more sustainable transport future. *Transport Reviews, 37*(6), 689–694.

Purvis, B., Mao, Y., & Robinson, D. (2018). Three pillars of sustainability: In search of conceptual origins. *Sustainability Science, 14*(3), 681–695.

Rebecchi, A., Boati, L., Oppio, A., Buffoli, M., & Capolongo, S. (2016). Measuring the expected increase in cycling in the city of Milan and evaluating the positive effects on the population's health status: A community-based urban planning experience. *Annali Di Igiene, 28*(6), 381–391.

Tsitoura, M., Tsoutsos, T., & Daras, T. (2014). Evaluation of comfort conditions in urban open spaces. Application in the island of Crete. *Energy Conversion and Management, 86*, 250–258.

Tsoutsos, T. (2018). Sustainable energy and mobility as drivers for the economic growth in the Mediterranean islands. *Renewable Energy and Sustainable Development, 4*(2), 56.

UN DESA. United Nations Department of Economic and Social Affairs. (2019). *68% of the world population projected to live in urban areas by 2050, says UN. UN DESA. United Nations Department of Economic and Social Affairs*. Retrieved August 20, 2019, from https://www.un.org/development/desa/en/news/population/2018-revision-of-world-urbanization-prospects.html.

Van Duppen, J., & Spierings, B. (2013). Retracing trajectories: The embodied experience of cycling, urban sensescapes and the commute between 'neighbourhood' and 'city' in Utrecht, NL. *Journal of Transport Geography, 30*, 234–243.

Van Holle, V., Van Cauwenberg, J., Deforche, B., Goubert, L., Maes, L., Nasar, J., Van de Weghe, N., Salmon, J., & De Bourdeaudhuij, I. (2014). Environmental invitingness for transport-related cycling in middle-aged adults: A proof of concept study using photographs. *Transportation Research Part A: Policy and Practice, 69*, 432–446.

WHO. (2004). *The global strategy on diet, physical activity and health (DPAS).* Retrieved August 20, 2019, from https://www.who.int/nmh/wha/59/dpas/en/.

WHO. (2010). *Global recommendations on physical activity for health.* Retrieved August 20, 2019, from https://www.who.int/dietphysicalactivity/factsheet_recommendations/en/.

Worldbank (2014). *CO2 emissions from transport (% of total fuel combustion). Data.* Retrieved August 20, 2019, from https://data.worldbank.org/indicator/EN.CO2.TRAN.ZS.

Worldbank. (2018). *Urban population (% of total population). Data.* Retrieved August 20, 2019, from https://data.worldbank.org/indicator/SP.URB.TOTL.IN.ZS.

Xu, S. J., & Chow, J. Y. Y. (2020) A longitudinal study of bike infrastructure impact on bikesharing system performance in New York City. *International Journal of Sustainable Transportation.*

Promoting Sustainable Mobility in Tourist Destinations: Mobility Center 2.0

Ingrid Briesner

Abstract Over the past decades, leisure-related mobility in European regions has increased continuously, especially in tourist destinations. New mobility patterns put enormous strain on sustainability issues in tourist regions, which are particularly vulnerable in this regard since the amount of individual mobility often is higher than in non-tourist regions leading to road congestion, seasonal changes of transport demand causing capacity problems, and high level of private car use increasing the need for parking spaces, etc. The rising importance of ecological tourism demands new perspectives of the tourist destinations in establishing new sustainable mobility structures and strategies for supporting regional economic development. Mobility Centers 2.0 are an efficient tool to reduce individual car use and the negative impact of visitor's travel in tourist regions, as well as to upgrade the quality of the leisure offer and the external image of the region. They can help to strengthen tourist regions as growth poles and improve the economic vitality of the targeted destination.

Keywords Mobility center · Mobility management · Mobility behaviour · Multi-modal information · Accessibility · Sustainable tourism · Eco-tourism · Cooperation · Synergies

1 Introduction

"We are millennials and we are looking to have different travel experiences than our parents" says Patrick Quayle, Vice President for international planning at United Airlines.

According to Booking.com (2019), "over half (55%) of global travellers are being more determined to make sustainable travel choices than they were a year ago, but barriers include a lack of knowledge and available or appealing options when trying to put this into practice". When it comes to in-destination experiences, over half

I. Briesner (✉)
IBCT -, Business Consulting & Training, Graz, Austria
e-mail: consulting@briesner.eu

© The Author(s) 2022
T. Tsoutsos (ed.), *Sustainable Mobility for Island Destinations*,
https://doi.org/10.1007/978-3-030-73715-3_7

(52%) of global travelers say they now alter behaviors to be more sustainable while traveling, such as walking, riding a bike, or hiking whenever possible.

Discount airlines and self-booking accommodation platforms have led to increased tourist flows in parallel with rising societal demand for more energy-efficient transportation. Such flows have created increased traffic congestion, less availability of public space due to increased demand for parking areas and higher use and circulation of private cars. The high motorization rate among tourists is a direct consequence of the lack of integrated public transport, tickets fit for visitors, insufficient provision of public transport connections to airports/harbors and leisure attractions (beaches, nature parks, golf courts, historic monuments, etc.), limited network perception, fragmented mobility information (which usually does not have the appropriate design and contents to ease usage), a lack of facilities and safety elements to ride bicycles or even walk in urban and interurban areas.

Changing demographics with an aging population put additional pressure on tourist destinations to create the proper precondition for the regions' future competitiveness, as the above-mentioned comfort requirements for tourists become more evident with age. The variability of touristic flows in terms of seasonality and space usage are in competition with the local resident supply of transportation and services, the local transport network and the local area demographic and social characteristics.

It is evident that inherent difficulty of small communities for the provision of integrated efficient transport services to residents and tourists often leads to higher individual car usage in order to compensate for the lack of appropriate mobility information and alternatives. Mobility Centers 2.0 are an efficient tool to reduce individual car use and the negative impact of visitor's travel in tourist regions, as well as to upgrade the quality of the leisure offer and the external image of the region. They can help to strengthen tourist regions as growth poles, which in turn improve the economic vitality of the targeted areas for both citizens and tourists, employees and employers, men and women.

2 Mobility Centers: A One-Stop Shop for Sustainable Mobility

Apart from the choice of transport mode to reach a tourist destination, the smart choice on-site plays a crucial role. Car rental businesses welcome the visitors already at the airport arrival halls and although a growing number of travelers become more and more aware of the necessity of sustainable, climate-friendly mobility, most of the tourist kilometers on-site are still done by car. In fact, the UN World Tourism Organisation (2008) states that climatically sustainable tourism requires fundamental shifts in consumer behaviour but the most probable explanation for the current mobility behaviour of tourists and visitors is the lack of information in unfamiliar settings.

Moreover, the still ongoing economic crisis and the growing trend of ecological tourism are demanding new integral strategies in a highly competitive field. Tourist destinations require solid and sustainable mobility structures, which can effectively support regional economic growth, as well as the development of safe, livable, and attractive places for all.

This is, where a Mobility Center 2.0–tailor-made for tourist destinations—steps in, and can serve as an information hub for all issues related to sustainable mobility in a region. All over Europe, Mobility Centers serve as information platforms, shaped after a model of a one-stop shop and their core business is multi-modal mobility information and advice (Fig. 1). Mobility centers are service facilities that offer users and potential users of public transport, information and services on the subject of mobility across all modes of transport. Ideally, they are the contact point for all questions about mobility.

The core business of the mobility centers is, in addition to multi-operational timetable information and ticket sales, to advise customers. Organizational services such as vehicle rental, car sharing, the sale of accessories, or the disposition of flexible transportation modes are also offered. Apparently, linking mobility services with leisure, cultural, and tourist information is becoming increasingly important. Tourism and leisure mobility requires a high degree of information about alternatives to the car. Successful examples for linking tourist information and mobility services are, e.g., the mobility center "Mobilito", Bischofshofen (AT), the "Mobicenter" Wuppertal (GER), or the "M.Punkt" in Wolfsburg (GER).

The special importance of the mobility center arises from the offer described above, combined with the simplicity of the system (one-stop shop) and the customer-friendly opening times. Through their work, mobility centers ensure that users of public transport feel very well looked after and are thus able to satisfy their mobility needs even without their own car. Especially in combination or cooperation with tourist information points, mobility centers can make a major contribution to the environmentally friendly handling of mobility needs in sensitive areas.

The Mobility centers which increasingly popped-up all over Europe differ significantly in terms of the range of services, external image, sponsorship, and financing. There are no specific networks and standards are usually not defined. This leads to a certain confusion of the offer and a lack of transparency for the user. Standardization, however, is to be seen as a prerequisite and is therefore essential, last but not least, to clearly communicate to the customer what services can be expected from a mobility center.

The minimum standards concern:

- Offers and work content
- Accessibility and opening times
- Personnel qualifications
- Equipment

The Mobility Center 2.0 expertise should at least cover:

- Information and advice on all modes of transport (public transport, on-demand services, e-mobility, walking, cycling)

Fig. 1 Mobility centers, established in the frame of SEE MMS project, 2012; South East Europe
Transnational Cooperation Programme

- Regional/national ticket sales for public transport, taking into account the entire service chain: information—advice—sales
- Advice on further alternatives to motorized vehicles
- Idea and complaint management
- Tourist information (guided tours, day-trips, landmarks) and ticket sales
- Organization or booking or information on demand-driven transports

The expansion stages of the mobility center relate to Mobility Management advice for schools or companies, cooperation with accommodation providers, the rental business, the implementation of a mobility shop, or even transport planning activities.

3 The Five Steps of Implementation

When implementing a Mobility Center, it is advisable to follow the five steps approach, as shown in Fig. 2:

Step 1: Involving all stakeholders from the start.

A diverse group of stakeholders has to be brought together for the implementation of a mobility center. Usually, transport operators, politicians, tourism institutions, residents, and lobby groups do not have a common view, which—on the other hand—is a pre-requisite for a tailor-made mobility center. The floor for successful implementation and long-term operation can be prepared through workshops and discussion rounds with key stakeholders, politicians, and all potentially involved people. This is where the foundation stone for the Mobility Center is laid and a tailor-made concept can jointly be developed, based on best practice in Europe. Besides the services to be

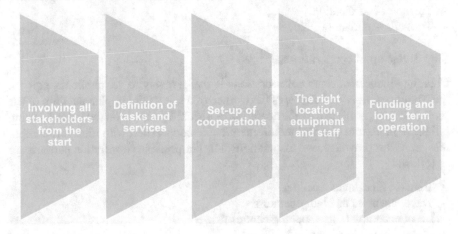

Fig. 2 The five steps of implementation of a mobility center

offered and the technical needs, also possible cooperations, expectations, strengths, and weaknesses should be investigated and discussed in a joint effort.

Step 2: Definition of tasks and services.

Mobility Centers offer information and service tasks for customers and represent customers' mobility interests. The target group includes all users of public transport, but also people who do not yet use public transport and who could be considered potential future customers. Other target groups, such as tourism, schools, and businesses, health organizations are increasingly being considered in tailor-made concepts. The following tasks and services are characteristic of Mobility Centers:

- General information on the transport network (routes, destinations, timetables, etc.)
- Route planning.
- Information on tickets and special offers.
- Ticket service (sales and/or reservation).
- Tourist Information.
- Information on car-sharing, car-pooling, e-moblity, and other regional offers.
- Bike rental.
- Information on bike rental (rental options available and/or tariffs).
- Information on pedestrian routes.
- Information on Park & Ride spaces.
- Complaints management.
- Additional services of a Mobility Center could include:
- Individual marketing for public transport.
- Project/event implementation.
- Organization of tourist tours.
- Planning and implementation of measures in the field of mobility management.
- Transport coordination.
- Traffic/urban planning, etc.

Step 3: Set-up of cooperations.

Creating alliances or setup collaborations is the best way to get public support for the Mobility Center and increases the chance of a successful implementation of the project. It is advisable to base the operation of a Mobility Center on several supporting pillars. A minimum standard requirement is the cooperation with transport companies, transport providers, and associations. Further possible cooperation partners are as follows:

- Tourist offices/municipalities
- Travel agencies and tour operators
- Local economy (e.g., shopping centers)
- Chambers of Commerce
- Employers' organizations
- Lobby groups (e.g., Cyclist Federations)

- Organizer of events and meetings
- Delivery or rental services, etc.

Step 4: The right location, equipment, and staff.

The location of a Mobility Center is crucial. It should be located in the city center or at a public transport hub to ensure the best possible presence and accessibility.

Another critical point in the implementation and organization of a Mobility Center is the personnel. Regarding personnel requirements, one can at least identify two different profiles, which are a manager on the policy level and the staff on the operational level. The task of mobility managers involves promoting the Mobility Center on the public and political levels. The mobility manager is in charge of the overall coordination, administration, team leading, and daily management of the information hub, as well as for the further development of the mobility center and the implementation of new mobility services. The staff´s tasks on the operational level are provision of information by phone, in writing and personal, as well as advice regarding all issues of mobility and travel awareness, Europe-wide ticket sale, planning of trips, bike rental, CarSharing—advisory service, management of information material, regular participation in team meetings, and general office work.

The standard equipment in a Mobility Center should be up to date, professional, inviting, and customer friendly. Concerning the technical facilities, computer hardware and software, a modern telephone system, a high-speed internet access, e-mail, and fax should be available. In times of digitalization, its services should be accessible 24 h online via Internet.

In case of further requirements for the stage of expansion of the Mobility Center or in order to create higher customer satisfaction, there are almost no limitations. However, a more sophisticated offer is also connected with higher costs and should be considered carefully.

Step 5: Funding and long-term operation.

The financial means are always scarce. Therefore, it is very important to consider the financial aspects in a realistic way and look out for co-operations. With regard to the time perspective for financing mobility centers, a distinction can be made between the introduction stage and the stage of running the Mobility Center in the longer term:

(a)   Investments and initial costs

Mobility Centers include the concept study, initial costs for the equipment (hardware, software, office equipment, mobile information facilities, etc.), training costs, and marketing (opening event).

(b)   Running Costs

The running costs include the rent and running overhead expenses (maintenance, communication, office equipment), costs for service staff and management, IT costs (Software, etc.), IT maintenance, office equipment, telephone expenses, office supplies, postal charges and marketing and costs for the conception and the production of information and publicity material as well as costs for advertisements and public awareness campaigns/publicity campaigns;

4   Cost/Benefit Analysis

In tourist regions, a Mobility Center has to take into account the seasonal and temporary needs of tourists and visitors. Those additional offers and services may lead to additional operational costs, but in fact, the combination of tourism and mobility provides valuable synergies, not only by building up on existing structures like, e.g., tourist information centers. Such synergies are immanent and can reduce the running costs, which are crucial for a long-term operation.

In terms of the cost/benefits, the traditional economic ROI evaluation methods are not found to be adequate for evaluating Mobility Centers based on a pure business enterprise selling goods and services. If these methods are used, it may not be possible to prove that the Mobility Center is "a profitable company". On the other hand, the cost/benefit analysis should be used to evaluate the Mobility Center as a "public service provider" and to consider the quantification and monetization of the social benefit in a more complex way.

An action-theoretical model for the choice of means of transport using the example of the mobility center in Graz (Weiss, 2004) analyzed the benefits of MOBIL ZENTRAL—the 1st Austrian Mobility Center in Graz, with a sample of 230 people, reliably proved the monetary advantages for the Styrian transport association (from the increased sale of tickets) after using the offers of the Graz Mobility Center. This advantage was put at 459.984 Euros through additional ticket sales per year and an overall change in behaviour, which would not have been realized without the work of the Mobility Center.

Other, rather social benefits (e.g., less noise, less road accidents, higher property values, etc.) were not included in this analysis. Also, it has to be taken into consideration, that the commitment to sustainable mobility in tourism, addresses a continuously growing group of tourists that is willing to travel the region sustainably. This results in growing numbers of overnight stays has the potential to boost the local and regional economy, and should be further analyzed.

5   Conclusions and Recommendations

The role of sustainable transport and mobility in the development of sustainable tourism is increasing, as leisure-time traffic is on the rise, and contributes considerably to greenhouse gas emissions, pollution, and climate change (DestiNet Services, 2019).

While arrival, accommodation and departure can be easily planned and booked via online platforms, the information to cover on-site needs and preferences of tourists and visitors, e.g., distribution in time and space, comfort requirements (carrying luggage, families with children, etc.), familiarity with the region's transport network, number of persons traveling together, purchasing power, language barriers, and special mobility information requirements, etc., is not covered online. The lack of multimodal on-site information is one of the main obstacles for tourists and visitors too, when trying to put their wish for sustainable leisure trips into practice.

Mobility centers providing high-quality information and advisory services on environmentally friendly on-site mobility, as well as promoting environmentally-friendly packages are an efficient tool to change the traveler´s point of view on a destination and can create a competitive advantage for tourist regions.

A main barrier in the implementation and long-term operation of mobility centers is the financing. Since the cost /benefit analysis in the traditional economic sense may not be applicable, the Mobility Center should be seen as a "public service provider" and the financial basis should include as many financing bodies as feasible. Particularly, in regions with high unemployment or a lack of alternative income, it may appear politically justified to use public funds in a way that people are kept in work or are given additional work. An alternative option to secure the operation of a mobility center is to tie it to existing regional-local structures and to integrate it into an overall organizational context.

One important success factor is that the service of mobility center for car-free tourism is communicated under the sustainability aspect, and should thereby focus on the additional benefits of car-free holidays and the additional relaxation that comes along with it. Sustainability as a unique selling proposition cannot create the product value that is worth buying.

If tourists or visitors are able to perceive the personal benefit of a car-free holiday, the journey is already booked and the indirect profitability inevitably will lead to economic success for the tourist region.

References

Booking.com. (2019). *Booking.com Sustainable Travel Report 2019*. Retrieved August 20, 2020 from https://globalnews.booking.com/bookingcom-reveals-key-findings-from-its-2019-sustainable-travel-report.

DestiNet Services (2019). *Tourism 2030: Travel, transport & mobility*. Retrieved Accessed 20, 2020 from https://destinet.eu/topics/sustainable-transport

UN World Tourism Organisation. (2008) *Climate change and tourism*. Retrieved August 20, 2020 from https://webunwto.s3-eu-west-1.amazonaws.com/imported_images/30875/climate2008.pdf

Weiss, E. (2004). Ein handlungstheoretisches Modell zur Verkehrsmittelwahl am Beispiel der Mobilitätszentrale in Graz, supported by Univ.- Prof. Dr. Karl Steininger, Inst. VWL, KFU Graz

Climate Change Mitigation Through Smart Environmental Monitoring of the Urban Ecosystem in Insular Touristic Cities: Experience from Rethymno and Madeira

Maria Aryblia, Lúcio Quintal, Miguel Ribeiro, Nikolaos Sifakis, Stavroula Tournaki, and Theocharis Tsoutsos

Abstract Cities, and general urban areas, contribute critically to climate change because of the GHGs related to traffic congestion, fossil fuel consumption, noise and air pollution. Air pollution, despite the noteworthy improvements during the last decades, still plays a significant role in the quality of living in European cities as it causes damages to health and ecosystems, thus making urgent the immoderate need of diminishing it. Within the Horizon 2020 CIVITAS DESTINATIONS project, six European island cities, which are highly attractive destinations for tourists worldwide, implemented various sustainable mobility measures. Among them, Funchal, Madeira-Portugal, and Rethymno, Crete-Greece, implemented two different environmental monitoring systems to measure and calculate a predetermined set of indicators, capable of estimating the environmental benefits in transport, economy, society, energy, and the environment. The monitoring systems were installed in specific sites all around these two cities, aiming, through the integrated sensors, to collect environmental data related to transport load, such as environmental indexes (temperature, humidity, noise) and air pollutants (CO_2, CO, NO_x, SO_2, PM). The collection of critical and reliable data offers the opportunity for an effective evaluation of the overall performance of the implemented measures toward sustainable, environmentally friendly, and low-carbon mobility policies.

Keywords Air quality · Transport impacts · Environmental impact assessment · Environmental monitoring · Urban environment

M. Aryblia · N. Sifakis · S. Tournaki · T. Tsoutsos (✉)
Renewable and Sustainable Energy Systems Lab, School of Chemical and Environmental Engineering, Technical University of Crete, Chania, Greece
e-mail: theocharis.tsoutsos@enveng.tuc.gr

L. Quintal
Arditi, Funchal, Portugal

M. Ribeiro
Computer Science and Engineering, Técnico Lisboa, Lisboa, Portugal

© The Author(s) 2022
T. Tsoutsos (ed.), *Sustainable Mobility for Island Destinations*,
https://doi.org/10.1007/978-3-030-73715-3_8

1   Introduction

Although significant changes over the past decades, air pollution remains a source of harm to human health and ecosystems (Park, 2020; Santos, 2017). Road transport, which contributes to a substantial share of Greenhouse Gas (GHG) transport pollution, plays a vital role in this, which significantly impacts air quality in urban spaces with superior road traffic levels (Santos, 2017). Road transport nowadays accounts for more than 82% of the personal transportation of people, including automobiles, two-wheeled vehicles, and buses/coaches. In comparison, 49% of freight traffic is road-related. Road transport vehicles often use fossil fuels, which emit large amounts of Greenhouse Gases (GHGs), and something noteworthy is that almost two-thirds of these pollutants come from light-duty automobiles (Bharadwaj et al., 2017; European Commission, 2017).

Road transport is the second-highest energy-consuming industry in Europe, after the industrial sector, and accounts for about 73% of overall GHG emissions from transport (Letnik et al., 2018). Air pollution often impacts other industries, including human health burdens, such as the industry, when job efficiency decreases under polluted conditions. As a consequence, sustainable mobility has been one of the main drivers for meeting the targets of the Paris Climate Change Agreement commitments. The adverse effects on the climate, energy, health, and economy of contemporary modes of transport have contributed to policies encouraging a sustainable future, which means cleaner, more secure, and more efficient mobility. In terms of fuel use, congestion, air and noise pollution, sustainability, connectivity, and quality of life, European cities face critical challenges (Andrés & Padilla, 2018; Santos, 2017).

It has also been shown that the environmental impacts of air pollution have a direct relationship with some severe health problems and are a vital risk factor for premature mortality (Brook et al., 2010; Fuks et al., 2017). Consequently, there is an urgent need to monitor the environmental condition of cities, especially the high traffic congestion areas, where there is increased air pollution, and therefore health risks (Park, 2020; Shekarrizfard et al., 2020; Wan et al., 2020). Due to the emerging importance of ensuring both human and ecosystems' health, several environmental monitoring systems have been implemented across the globe, and various research works have been conducted over the last decades regarding this specific subject (Andersson et al., 2020; Tezel et al., 2019). These systems constitute a useful tool to provide real-time monitoring of environmental health conditions and help to prevent serious problems. Besides, they are a handy tool for delivering valuable data that can be utilized to measure various specific indexes and evaluate the effectiveness of measures regarding sustainable mobility, such as those implemented in the CIVITAS DESTINATIONS initiative (Tournaki et al., 2017, 2018).

To overcome the problems mentioned above and achieve a balanced low-carbon economy by 2050, Europe is seeking to change the need to reduce transport pollution by at least 60% relative to 1990. The incremental shift away from fossil fuel-based transport systems and the adoption of new and innovative low-carbon solutions was already an objective in the Transport White Paper of 2011. It was outlined in various policy documents (European Commission, 2003, 2017). In particular, some of the measures taken are to reduce the speed limit within cities to force citizens to make more use of public transport. Moreover, a more successful approach will be to outlaw the most polluting cars and place substantial penalties on offenders. Therefore, a monitoring scheme for air emissions, car traffic control, and air quality documentation is required to allow proper use of these data and improve the situation. A noteworthy alternative regarding sustainable mobility is the use of alternative fuels, such as biodiesel, in some cases with several significant advantages and weaknesses (Foteinis et al., 2018; Tsoutsos et al., 2010, 2019). An additional alternative that enhances the sustainable mobility concept refers to the use of Electric Vehicles (EVs) or Hydrogen-Electric Vehicles (HEVs), which both seem to play a vital and determinant role in the future of the transportation sector (Pavić et al., 2020; Smaragdakis et al., 2020).

Six European island cities, attracting a high number of tourists from Europe and worldwide, are implementing sustainable mobility measures within the Horizon 2020 CIVITAS DESTINATIONS project; Funchal (Madeira-Portugal), Las Palmas de Gran Canaria (Spain), Elba (Italy), Valletta (Malta), Limassol (Cyprus), and Rethymno (Crete-Greece). A refined Impact Assessment Framework has been developed, which includes a set of indicators to estimate the environmental benefits in transport, economy, society, energy, and the environment, due to the DESTINATIONS measures. This assessment methodology is based on the guidelines of CIVITAS SATELLITE, adapted to project objectives and specific insular conditions, with the addition of new indicators in terms of pollution and energy (Tournaki et al., 2017, 2018). In this book chapter, the two representative environmental monitoring cases in Madeira and Rethymno are introduced to strengthen the mobility policy and preparation.

After the introductory section, a thorough insight of the two environmental monitoring systems is presented, and at last, the conclusions regarding the effectiveness and the applicability of these two systems are being evaluated and demonstrated.

2 A Detailed Description of the Two Systems

2.1 The Case of Rethymno

A methodology was developed and applied in Rethymno under the CIVITAS DESTINATIONS initiative, which continuously monitors the acquired data during different annual periods, integrating the seasonal fluctuation due to tourism, outdoor activities, and festivals (highly crowded events).

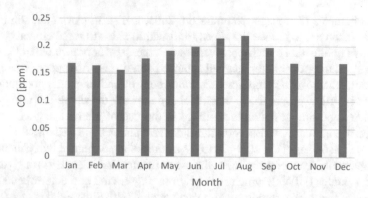

Fig. 1 Seasonal fluctuation of CO for the city of Rethymno

2.1.1 Overview of the Area of Study (Rethymno)

Rethymno is the third biggest city in the island of Crete, with a total area of 397.48 km^2, and a total population of 63,000 inhabitants. Indicatively, it has been calculated that the index describing the number of vehicles per 1,000 residents equals 533, which means that at least every household in the city owns a vehicle. The town, on the contrary, does only have a port that serves one or two times a week, during summer months, some itineraries to other Greek islands. The main airports of the island are located in Heraklion (73% of total arrivals) and Chania (27% of total arrivals). The main entrances for the city are the island's highway 90 and its port.

Almost 520,000 tourists yearly are being accommodated in municipality's hotels, and it is estimated that more than 30% of these tourists are staying in suburbs, being served by the city's Public Transport (PT)/mobility services, and more than 1.5 million tourists visit the city on day tours and cruises per year. Besides, let alone the permanent residents of the island, during summer months, the environmental condition of the town is considerably burdened by tourism, and the seasonality fluctuation is very intense. Consequently, the data acquired from the implemented systems evaluate and examine the influence of the flux of tourism. Indicatively, the seasonal fluctuation of CO is depicted in Fig. 1.

2.1.2 Insight in the Pilot Environmental Monitoring System

Overview

The monitoring system in Rethymno, developed by the Renewable and Sustainable Energy Systems Lab of the Technical University of Crete (ReSEL-TUC), is a low-cost environmental monitoring equipment designed as a part of the CIVITAS DESTINATIONS project. The system employs a two-level scheme, at pedestrian and building level, shaping a two-level monitoring system, recording and comparing pollutant concentrations during rush hours. Three meteorological stations, with integrated environmental monitoring sensors, are installed on buildings within the city, conducting measurements of air pollutant concentrations, noise level, ambient air temperature, humidity levels, and wind speed on a 5-min interval. The recorded measurements are stored in a database. They can be presented online in real-time via webpage with a modification on the software algorithm and the installation of a SIM card on the 3G modem of the device. The above stations are non-intrusive (above ground) and consist of: (a) support base, (b) specially designed housing, (c) integrated sensors, and (d) a recording unit and data storage and transmitting software.

The sensors, as well as the unit for recording and sending data, are integrated into a standard shell that is compact and made of such materials to ensure both protection from external weather conditions and its smooth operation throughout the lifetime of the system. The sensors were selected and appropriately integrated into the system so that the software can detect each pollutant individually, and a cross-sensitivity case is excluded. Each measurement is unique and captured as accurately as possible in the actual situation. For data collection, the sensors function as stand-alone systems and are connected through the unique software to the host computer, where the data is recorded, and stored.

Technical specifications

The environmental station is an autonomous, fixed mounting air quality monitoring station that tracks the key air pollutants and weather parameters. The equipment is placed into urban areas of cities to collect localized data for air pollution, noise pollution, and microclimate conditions. The sensors and the electronic circuits are protected against the weather using an industrial enclosure with column mounting support. The data are stored into the included microSD card. They can be transferred in real time via the **GPRS** network to the cloud to be stored and used for the depiction of the environmental conditions all over the city if a SIM card is inserted.

It features a low power and high accuracy real-time clock with a battery-backed based on the DS3231. On the start-up, the station initializes a local Wi-Fi network, which allows the authenticated users to download the historical data directly to their connected devices. Furthermore, the **GSM** module establishes a connection to the remote server through GPRS and push new data in fixed 5-min intervals using the **MQTT** protocol. Figure 2 illustrates the hardware architecture of the prototype implementation. The selected **MCU** includes multiple hardware interfaces capable of maintaining the connection with all sensors and modules simultaneously.

The air pollution measurements are acquired through a high-quality scattering sensor to detect the Particulate Matter PM2.5/PM10 concentration in the air. Additionally, the prototype contains the MiCS-6814, which is a compact Metal Oxide Semiconductor (**MOS**) sensor with three fully independent sensing elements to measure concentration levels of pollutants such as carbon monoxide; nitrogen dioxide, and ammonia. This configuration includes a high-precision non-dispersive infrared sensor to measure the concentration levels of carbon dioxide in the air. The technical specifications of the air pollution sensors are presented in Table 1.

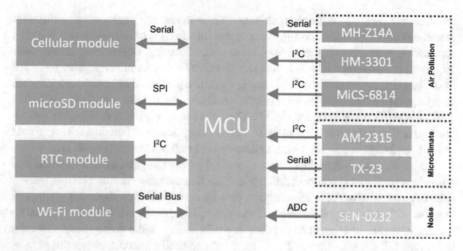

Fig. 2 Hardware architecture and interfaces

Table 1 Technical specifications of air pollution sensors

Sensor	Parameter	Range	Resolution	Accuracy	Lifespan
MiCS-6814	CO	0 to 1000 ppm	0.1 ppm	0.5 ppm	>5 years
	NO_2	0 to 10 ppm	20 ppb	5%	>5 years
	H_2	0 to 1000 ppm	1 ppm	5%	>5 years
	NH_3	1 to 500 ppm	1 ppm	5%	>5 years
	C_2H_5OH	0 to 500 ppm	1 ppm	5%	>5 years
MH-Z14A	CO_2	0 to 5000 ppm	50 ppm	3%	>3 years
HM3301	Particle matter	1 to 500 ug/m^3	1 ug/m^3	5%	>2 years

For the microclimate measurements, the AM-2315 digital temperature and humidity sensors are used. On the top of the enclosure is placed a TECHNOLINE TX 23 wind sensor capable of measuring high wind speeds up to 180 km/h. Furthermore, this sensor includes a wind direction sensor with a high-resolution digital decoder. The technical specifications of the microclimate sensors are presented in Table 2.

Furthermore, for the noise pollution, an instrument circuit and low noise microphone to accurately measure the sound level of the surrounding environment are

Table 2 Technical specifications of the microclimate sensors

Sensor	Parameter	Range	Resolution	Accuracy	Lifespan
AM-2315	Temperature	−20 to 80 °C	0.1 °C	0.1 °C	>10 years
	Relative humidity	0 to 100%	1%	2%	> 10 years
TX-23	Wind speed	0 to 50 m/s	0.1 m/s	0.9 m/s	>10 years
	Wind direction	0 to 360°	22.5°	22.5°	>10 years

featured. This circuit provides linearity between the output voltage and the decibel value, which leads to a simple conversion equation avoiding the logarithmic computations in the **Micro-Controller Unit (MCU)**. The technical specifications of the noise sensors are presented in Table 3.

Lastly, the general technical specifications of the installed stations are presented in Table 4.

Table 3 Technical specifications of the noise sensor

Sensor	Parameter	Range	Resolution	Accuracy	Life span
SEN-0232	Sound/Noise	30 to 130 dB	1 dB	1.5 dB	>10 years

Table 4 General technical specifications of the stations

Category	Parameter	Value
General	Dimensions	135 × 170 × 441 mm
	Weight	1.7 kg
	Power supply cable length	10 m
	Color	Light grey
Power supply	A/C input	240 VC, 50/60 Hz, 5A
	A/C output	5 V DC, 1A
	Weather resistance	IP 65
	Plug type	Type F
Wireless network	Connectivity	Wi-Fi 2.4 GHz
	Standards	IEEE 802.11b/g/n
	Frequencies	2400–2483.5 MHz
	Chip	ESP8266
	Certifications	CE, FCC
Cellular network	Connectivity	GPRS over GSM
	Frequencies	850, 900, 1800, 1900 MHz
	Chip	SIMCom SIM800L
	Certifications	CE, GCF, FCC, TA, CTA, RoHS
Internal storage	SD card	16 GB
	SD card format	FAT32

2.1.3 Exact Siting (Sites) of the Implemented Systems

The representatives for the city measurement locations were selected based on the open discussion with all the city's policy-making actors (municipality, tourism office, chamber of commerce, technical chamber, associations of citizens). The three meteorological stations, alongside with eleven thermocameras, are located in critical city points that present high traffic activity. Regarding the meteorological stations, they are equipped with durable sensors in building level height, to avoid missing data and damages. The exact location of each station is depicted in Fig. 3.

Stations 01, 03, and camera 05 are located next to the sea in the regional road of Rethymno near Fortezza Castle, where a significant number of cars are passing by each day, particularly during the summer months because of the attractiveness of the landmark. Station 03 and camera 06 are located in the south entrance to the city of Rethymno for passengers who are coming through highway 90. Besides, a market is set up every Tuesday near this site, and increased traffic load is observed during the operation hours. These two options are selected because there are both thermocameras and meteorological stations in these sites. Consequently, correlations between traffic (vehicle count) and several atmospheric pollutants can be drawn.

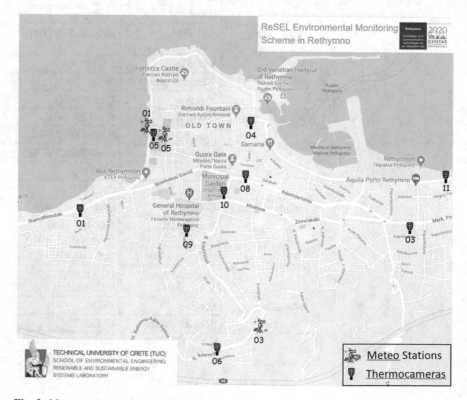

Fig. 3 Meteo stations and thermocameras sites

2.1.4 Indicative Outcomes of the Data Deriving from the System

The measurements are taken in fixed 5-min intervals. The data are accessible remotely using an API interface or directly through the local Wi-Fi network of the device. The purpose of this prototype implementation is to establish a global network of monitoring stations in the urban areas focused on continuous environmental observations. Combining the different connectivity options and low power consumption, this solution can be deployed for a large variety of field applications. The router has, after all, packages installed, adequate storage to save more than two years of data. In addition, the data is received by the Wi-Fi device or is sent to a central server in a proper format that can be easily analyzed by the receiver. Figure 4 depicts an example of the visualization of the data.

Due to the operation of the environmental monitoring stations and the acquired measurements, several conclusions regarding the seasonality of the environmental condition of the city can be drawn, such as that the concentration of CO_2 in the city is higher during winter months than during summer months. Many may argue with this due to the increased tourism factor. However, there is strong evidence that the use of heating equipment mainly causes this seasonality during winter, which emits a lot more CO_2 than cars and generally tourism activity. The aforementioned can be observed in Fig. 5, which depicts the daily mean concentration of CO_2, for both winter and summer.

By taking advantage of the acquired data from the thermocameras alongside the data of the environmental monitoring stations, there is an excellent bunch of correlations that can be made, and several conclusions can be drawn. Indicatively, the

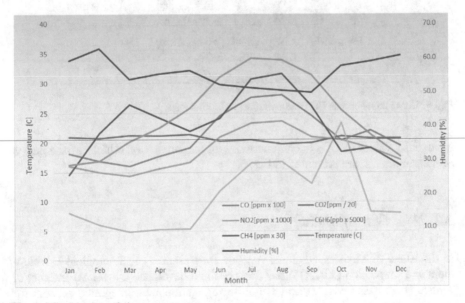

Fig. 4 Monthly data of air measurements

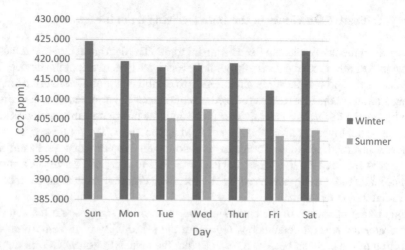

Fig. 5 Mean daily CO_2 concentration among the two seasons

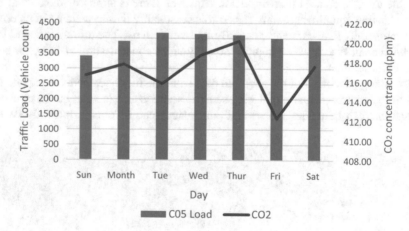

Fig. 6 Correlation between CO_2 concentration and traffic load

mean daily values of CO_2 in conjunction with the number of vehicles, for a period of measurements, are presented in Fig. 6.

2.2 The Case of Madeira

In the island of Madeira, Regional Agency for Research, Technological Development and Innovation (ARDITI) has designed and developed a low-cost Internet of Things (IoT)-based environmental monitoring station with integrated sensors for real-time measurement of CO_2, CO, NO_2, O_3, SO_2, NO, PM2.5/PM10, noise, temperature, humidity, and atmospheric pressure. The smart system is installed in eight urban locations in the city of Funchal in order to collect data related to transport load and the environmental conditions in the whole city. Three such stations are located in key entry spots in Funchal where the municipality also installed a traffic counting/monitoring system.

Gathered data can then be related to urban accessibility, traffic and environmental indicators, and users satisfaction. It makes use of smart, low-cost solutions, based on wireless networks and crowd-sourced data. The Environmental Monitoring Station presented here is one of the results of ARDITI work in CIVITAS DESTINATIONS project.

2.2.1 Hardware Overview

The sensors were selected based on their quality, cost, availability, and openness (to external hardware and software connection). Subsequently, the software was developed (mainly in Python) and the sensors were connected to the central unit (a smart router/embedded computer), which in the current version is the GL-AR300M Mini Smart Router, from GL-iNet manufacturer. Sensor control, reading, and communication software is installed and configured in the local embedded computer which runs the necessary software to control the sensors, measure indicators, and communicate with the server database. With the intent to make the system modular, we chose sensors that have a USB interface, or that could be read through intermediate hardware (see Table 5). We use a USB hub to accommodate the sensors required for each location, connected with the embedded system (in our case, using a router, we exploit its connectivity features to access real-time sensing data). The sensors that

Table 5 Technical specifications of sensors used (ARDITI, 2019)

Reading/Indicator	Sensor used	Output/Interface
Particles (PM2.5, PM10)	Nova PM sensor SDS011	USB
Ozone (O_3)	Sensoric O3 3E 1	4–20 mA
Carbon Dioxide (CO_2)	Yocto-CO_2	USB
Sulfur Dioxide (SO_2)	SO_2 SL 4–20 mA Transmitter	4–20 mA
Nitrogen Dioxide (NO_2)	NO_2 2E sensor with 4–20 mA transmitter, 0–20 ppm	4–20 mA
CO	CO ECO-Sure (2e) 4–20 mA Transmitter	4–20 mA
Nitrogen Oxide (NO)	4-NO-2000	0–5 V
Temperature, humidity, air pressure	Yocto-Meteo	USB
Noise	Digital microphone with integrated sound card	USB

have output in the current (mA) or voltage (V) are read by USB current and voltage sensors from Yoctopuce (Yocto-4-20mARx and Yocto-0-10 V-Rx).

2.2.2 Software Overview

The access to the embed system is done remotely, to transfer files and run system commands. The software to read the sensors (sensor poller) is in the embedded system and is responsible for capturing the data points and send them to a central server employing a network connection. The system is composed of a capturing component, an upload manager, and a data sender. The capturing component interacts with the different modules that acquire the data and forwards it to the upload manager. The upload manager checks the server connectivity and either sends the current and saved data, or stores it locally when no connection is found. The upload manager passes the data to the data sender, which authenticates the requests and actually sends them to the central server. The VPN connection can be used to remotely control the system and perform maintenance, or software upgrades, access the embed system, to transfer files and/or run system commands. Figure 7 illustrates the software architecture implemented.

The modular platform has been designed so as to integrate different sensors in order to satisfy the requirements of each city and also to accommodate different methods of data communication, ranging from the cabled network, Wi-Fi, USB 3G modem, or the possibility of LoRa (see Fig. 8). This gives the possibility for each location to have specific characteristics, where the number of sensors integrated into the device and the communication method can be selected/decided for each location.

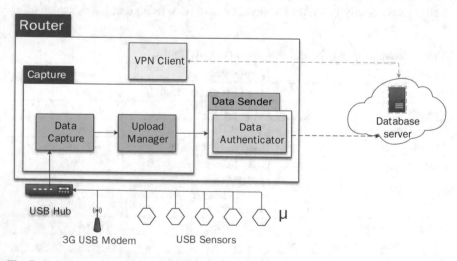

Fig. 7 Software architecture overview (ARDITI, 2019)

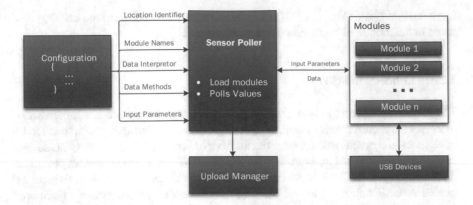

Fig. 8 Sensor poller overview

The collected data are presented in a real-time web page and stored in a central database for monitoring and historical recording. The overall infrastructure/system represents a smart, low-cost alternative solution for environmental monitoring, based on wireless networks and crowd-sourced data. If compared to existing commercial solutions, this solution offers a low-cost alternative to cities for air quality/noise monitoring. It can easily be adapted/customized for different scenarios. 3D model files for the casing have been created in open Drawing Interchange Format (DXF).

2.2.3 Prototype

With the interconnected sensors encased in a box to accommodate all the components and sensors, the boxes are not waterproof, due to the need of the exposure of the sensors to the outside air, thus needing protection shields as shown in Fig. 9. The protection is attached to the box and attached to a supporting mount. Three housing prototypes have been created and tested: one of such prototypes was used as the base to estimate the right dimensions and compartments of the final version of the

Fig. 9 Sensors and housing in the final prototype

box/housing. Then, using our 3D (DXF) model files, construction of the housing was made in "carbon steel" by a subcontracted provider.

2.2.4 Connection Scheme

The modular platform has been designed so as to integrate different sensors in order to satisfy the requirements of each city. This gives the possibility for each location to have specific characteristics, where the number of sensors integrated into the device can be selected/decided by each city. As presented in Fig. 10 some sensors provide a direct reading via USB while others require either a voltage or a current intermediate reader. A USB hub provides all the USB ports needed by the system, including connection to a 3G Modem in case Wi-Fi Internet is not available to the router (e.g., in remote zones). The only particular case to note is the NO sensor, which requires a voltage reader and power supply of 5 V. To achieve this, a DC/DC converter connected to the 23 V output of the voltage reader was used. Then an amplifier (OpAmp) was used to maintain the output signal readable. Figure 11 presents the connectors inside the housing/box and how it is all assembled in the final prototype.

Not considering human resources costs, 3,000€ was the estimated cost to purchase and assemble components, produce the custom case and assemble the current prototype of the station (concrete installations may incur in additional costs). The total monitoring system has been developed in order to be a transferable solution to support green mobility and have replicability potential; it is a low-cost solution, can be adopted and installed at any city, while ARDITI can provide the necessary support for installation (the list of components/sensors, 3D model files for the protective casing in DXF format and the software that needs to be installed in the embedded system, api, database, and data visualization).

2.2.5 Deployment

The sensor stations were installed in 8 urban locations in the city of Funchal, with three such stations located in key entry spots in Funchal where the municipality also installed a traffic counting/monitoring system. Figure 12 shows the location of the sensors throughout the city.

The data captured, if stored locally, can be retrieved via SFTP with a capacity for approximately 4 months of local data storage. If the data is sent to a central server, the API that receives it, can also provide API endpoints to visualize it, as is the case of the one implemented by ARDITI, where the data can be accessed via a web platform that accesses the API and displays the real-time data (Fig. 13).

The collection of critical and reliable data can lead to an adequate evaluation of the overall performance of implemented measures, toward sustainable, environmentally friendly and low-carbon mobility policies. The environmental monitoring schemes presented in this work can provide an efficient, reliable, replicable, low-cost, continuous view and data of the urban environment, allowing urban planners and authorities

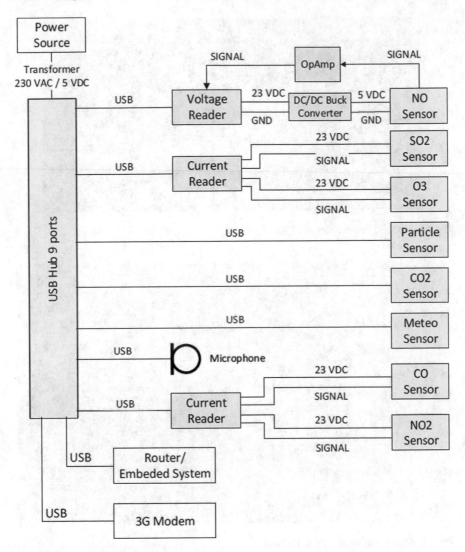

Fig. 10 Sensor connection scheme

to locate traffic congestion points and analyze the environmental impact of transportation; therefore, to adapt their mobility strategies and to formulate suitable sustainable mobility solutions if/when/where needed.

It is worth to notice that the authorities remain in charge to implement sustainable strategies and take decisions, but tools like these can support decision-making as they provide an easy way to visualize and detect relevant/not usual situations and to enhance awareness, supporting and facilitating the communication between the different community's needs. In fact, sustainable development requires to join the

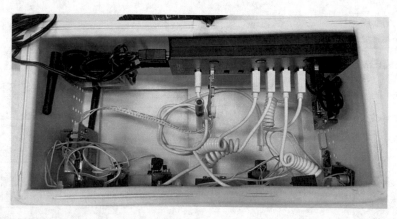

Fig. 11 Sensors, smart router and connectors

Fig. 12 Sensor map in the city of Funchal

effort of the different communities and authorities, to consider its social dimension, and to become an effective framework to improve sustainable mobility and tourism.

The collected sensing data can be put to good use to improve the daily mobility experience, employing sustainable mobility by monitoring its effects in the air quality of the cities. Moreover, this solution represents an opportunity to investigate the possibility to meet the different needs, by providing environmental-related information not only to the official entities, but also empowering the different communities, by collaborating in the sustainable development. To promote this idea, we plan to evolve our platform into a full community network capable of offering other personalized

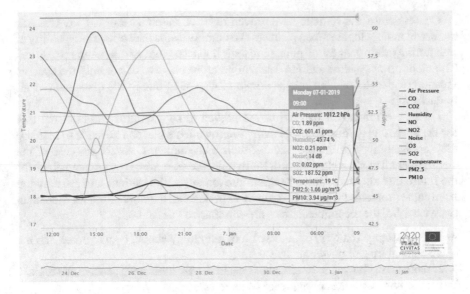

Fig. 13 Real-time sensor data visualization

services to communities, exploring crowdsourcing and crowdsensing where citizens can acquire their own open-source stations and contribute to the network.

3   Conclusions and Recommendations for Future Research

In this chapter, two novel low-cost IoT environmental monitoring systems were demonstrated for the evaluation of the environmental condition of two European cities, which have in common that during summer months, there is high touristic demand. Therefore, the air pollutants' levels are supposed to skyrocket. After the preliminary analysis of the acquired data, it was proven that this assumption is false, as the environmental condition of the cities seems to worsen, but still, the pollutants' concentrations do not increase that much to override the legislation limits and become dangerous for human health.

It is also shown that such systems can be an excellent evaluation tool for any sustainability measure in the vital aspect of environmental quality and air pollution, and therefore human and ecosystem's health. Rethymno is proven to be a pollution-free city even on periods of peak demand when the traffic congestion in the central city's streets is high, and the GHGs from the vehicles are increased. Even in periods where festivals and other highly crowded events take place, the city remains in average levels of air pollution, exhibiting some momentary measurements out of legislative limits.

Consequently, the stations have proven to be an essential tool for any research team, and mainly for the urban planners and authorities, as they offer the capability of monitoring traffic congestion points and their impact on the environmental condition. The monitoring systems and the data can be representative of not only the specific location of the stations but for the whole city, as well due to the dispersion of air pollutants.

Lastly, even if these two systems are proven to be a reliable and essential tool for the interested parties, there are many improvements that can be made to increase their applicability and the precision of their measurements. As for the case of the environmental monitoring system of Rethymno, there are some recommendations for the future that aim to the incremental improvement of the stations and are deriving from some problems that have risen up by experiencing their operation for more than two years. So, the recommendations are summarized as follows:

- Energy storage (battery) of at least 1 h of autonomy should be integrated into the stations in case of power outages;
- If more precision on measurements is desired, new innovative sensors can replace the old ones with the consequence of extra cost;
- The software's algorithm should be updated with a newly created script that could easily depict the measurements in easily comprehensible figures that will be easy to import in any of the authorities' website;
- Better wind and dust sensors should replace the already integrated due to their complex operation.

References

Andersson, E. M., Ögren, M., Molnár, P., et al. (2020). Road traffic noise, air pollution and cardiovascular events in a Swedish cohort. *Environmental Research, 185*, 109446. https://doi.org/10.1016/j.envres.2020.109446

Andrés, L., & Padilla, E. (2018). Driving factors of GHG emissions in the EU transport activity. *Transport Policy, 61*, 60–74. https://doi.org/10.1016/j.tranpol.2017.10.008

Bharadwaj, S., Ballare, S., & Rohit, C. M. K. (2017). Impact of congestion on greenhouse gas emissions for road transport in Mumbai metropolitan region. *Transportation Research Procedia, 25*, 3538–3551. https://doi.org/10.1016/j.trpro.2017.05.282

Brook, R. D., Rajagopalan, S., Pope, C. A., et al. (2010). Particulate matter air pollution and cardiovascular disease: An update to the scientific statement from the American heart association. *Circulation, 121*, 2331–2378. https://doi.org/10.1161/CIR.0b013e3181dbece1

European Commission. (2003). *The revised renewable energy directive.*

European Commission. (2017). An agenda for a socially fair transition towards clean, competitive and connected mobility for all. *Com, 177*, 18.

Foteinis, S., Antoniadis-Gavriil, A., & Tsoutsos, T. (2018). Life cycle assessment of algae-to-biodiesel shallow pond production systems in the Mediterranean: Influence of species, pond type, by(co)-product valorisation and electricity mix. *Biofuels, Bioproducts & Biorefining, 12*, 542–558. https://doi.org/10.1002/bbb.1871

Fuks, K. B., Weinmayr, G., Basagaña, X., et al. (2017). Long-term exposure to ambient air pollution and traffic noise and incident hypertension in seven cohorts of the European study of cohorts for air pollution effects (ESCAPE). *European Heart Journal, 38*, 983–990. https://doi.org/10.1093/eurheartj/ehw413

Letnik, T., Marksel, M., Luppino, G., et al. (2018). Review of policies and measures for sustainable and energy efficient urban transport. *Energy, 163*, 245–257. https://doi.org/10.1016/j.energy.2018. 08.096

Park, Y. M. (2020). Assessing personal exposure to traffic-related air pollution using individual travel-activity diary data and an on-road source air dispersion model. *Heal Place, 63*, 102351. https://doi.org/10.1016/j.healthplace.2020.102351

Pavić, I., Pandžić, H., & Capuder, T. (2020). Electric vehicle based smart e-mobility system—Definition and comparison to the existing concept. *Applied Energy, 272*, 115153. https://doi.org/ 10.1016/j.apenergy.2020.115153

Santos, G. (2017). Road transport and CO_2 emissions: What are the challenges? *Transport Policy, 59*, 71–74. https://doi.org/10.1016/j.tranpol.2017.06.007

Shekarrizfard, M., Minet, L., Miller, E., et al. (2020). Influence of travel behaviour and daily mobility on exposure to traffic-related air pollution. *Environmental Research, 184*, 109326. https://doi.org/ 10.1016/j.envres.2020.109326

Smaragdakis, A., Kamenopoulos, S., & Tsoutsos, T. (2020). How risky is the introduction of fuel cell electric vehicles in a Mediterranean town? *International Journal of Hydrogen Energy, 45*, 18075–18088. https://doi.org/10.1016/j.ijhydene.2020.04.224

Tezel, M. N., Sari, D., Ozkurt, N., & Keskin, S. S. (2019). Combined NOx and noise pollution from road traffic in Trabzon, Turkey. *Science of the Total Environment, 696*, 134044. https://doi.org/ 10.1016/j.scitotenv.2019.134044

Tournaki, S., Farmaki, E., Tsoutsos, T. (2018). Impact assessment of sustainable mobility in touristic cities of Europe: The CIVITAS DESTINATIONS approach on energy, environment and economy. *Green Energy Technology* 297–309. https://doi.org/10.1007/978-3-319-89845-2_21

Tournaki, S., Farmaki, E., & Tsoutsos, T. D. (2017). *Sustainable mobility solutions and impact assessment in touristic areas in CIVITAS DESTINATIONS project The case of Rethymno.*

Tsoutsos, T., Kouloumpis, V., Zafiris, T., & Foteinis, S. (2010). Life Cycle Assessment for biodiesel production under Greek climate conditions. *Journal of Cleaner Production, 18*, 328–335. https:// doi.org/10.1016/j.jclepro.2009.11.002

Tsoutsos, T., Tournaki, S., Gkouskos, Z., et al. (2019). Quality characteristics of biodiesel produced from used cooking oil in Southern Europe. *ChemEngineering, 3*, 19. https://doi.org/10.3390/che mengineering3010019

Wan, Y., Li, Y., Liu, C., & Li, Z. (2020). Is traffic accident related to air pollution? A case report from an island of Taihu Lake, China. *Atmospheric Pollution Research, 11*, 1028–1033. https:// doi.org/10.1016/j.apr.2020.02.018

Printed in the United States
by Baker & Taylor Publisher Services